WONDROUS HEALING

Wondrous Healing

Shamanism,

Human Evolution,

and the Origin of Religion

James McClenon

NORTHERN ILLINOIS UNIVERSITY PRESS / DEKALB

© 2002 by Northern Illinois University Press

Published by the Northern Illinois University Press,

DeKalb, Illinois 60115

Manufactured in the United States using acid-free paper

All Rights Reserved

Design by Julia Fauci

Library of Congress Cataloging-in-Publication Data

McClenon, James.

Wondrous healing: shamanism, human evolution,

and the origin of religion/James McClenon.

    p.   cm.

Includes bibliographical references and index.

ISBN 0-87580-284-2 (alk. paper)

1. Religion. 2. Shamanism. 3. Spiritual healing.

4. Human evolution--Religious aspect. I. Title.

BL430 .M33 2001

200--dc21

2001132327

Contents

Illustrations

Tables

Acknowledgments

Sections of this book contain copy from some of my previously published work. The article "Spiritual Healing and Folklore Research: Evaluating the Hypnosis/Placebo Theory" was published in *Alternative Therapies in Health and Medicine* (Vol. 3, no. 1 [1997]: 61–66). It appears in modified form in chapter 4 and is used with that journal's permission. The article "Content Analysis of an Anomalous Memorate Collection: Testing Hypotheses Regarding Universal Features" was published in *Sociology of Religion* (Vol. 61, no. 2 [2000]: 155–69). It appears in modified form in chapter 5 and is used with that journal's permission. Accounts from my book *Wondrous Events: Foundations of Religious Belief* (Copyright © 1994 by James McClenon) are published with permission of the University of Pennsylvania Press and appear in chapters 5 and 6. The article first describing the ritual healing theory, "Shamanic Healing, Human Evolution, and the Origin of Religion," was published in the *Journal for the Scientific Study of Religion* (Vol. 36, 1997: 345–54). Ideas and text presented in that article provide the foundation for this book and are used with the permission of that journal. Material about the "Baltimore Haunting" and the "Suicide Haunting" was previously published in McClenon, "The Sociological Investigation of Haunting Cases," chapter 4 in *Hauntings and Poltergeists: Multidisciplinary Perspectives*, ed. James Houran and Rense Lange (Jefferson, N.C.: McFarland, 2001) and is used with permission of McFarland and Co., Inc., Publishers.

I wish to thank Steve Marsh, Associate Professor of Literature, Elizabeth City State University (ECSU), for his comments pertaining to an earlier version of this manuscript; Harold W. Ellingsen, Jr., Assistant Professor of Mathematics, ECSU, who assisted with the mathematical models described in the appendix; Cheryl Sutton, Reference Library Assistant, ECSU, who helped locate references; and Diane Patterson, Graphics, ECSU, who aided in preparing charts for publication. Martin Johnson, Acquisitions Editor at Northern Illinois University Press, helped guide this project to completion. The National Archaeological Museum, Athens, Greece, granted permission to use photographs of Greek healing gods. The Fortean Picture Library, Ruthin, UK, granted permission to use a photograph of a psychokinesis "sitter" group.

WONDROUS HEALING

Introduction

Evolution and Religion

People have puzzled over the origins of religion for many centuries. One scenario suggests that the pre-linguistic ancestors of humans *(Homo erectus)* sang in a chanting manner while sitting around their fires. We might imagine the scene: a creative female devises repetitive songs that the group adapts. She sings with a clear, well-controlled voice, and many males are attracted to her. She selects a skilled male singer for her mate, and her children are able to sing even more clearly. She is part of a process leading to the development of language.

The hominids stare at the fire, chanting the hypnotic, wordless songs together. As they sing, two sick children lie on the ground, waiting with anticipation. The creative female strongly hopes they will recover. She closes her eyes and feels caught up by the chanting. For her, normal life has disappeared. She has been taken to another world, a world where she has joined the fire. She rises, dances toward the coals, and walks safely over the embers. The group is transfixed by her behavior. She feels a powerful sensation flow through her hands as she touches the sick children. They know she is trying to help them.

One child completely relaxes, his eyelids fluttering. He feels a powerful sensation coming from her hands. By the morning, his fever has declined. But the other child remains frightened, and a few days later he dies. To the degree that their mental abilities allow it, the hominids ponder these events: How did the female dance safely over the hot coals? Why did one child die while the other was healed?

The Ritual Healing Theory

This wondrous healing ritual, and others like it, contributed to the beginning of religion. A series of stages led to the modern biological capacity for religious belief. Ancient primates practiced rudimentary rituals, but the "healing"

that each derived from these activities varied. Some of the primates benefited more than others and, as a result, possessed survival advantages.
Many rituals produce trances, hallucinations, and other altered states of
consciousness due to their repetitive, hypnotic qualities. In our story, the
Homo erectus child who was more suggestible recovered, grew up, and
passed on his genes to his children. The child who was not suggestible did
not survive. The result was evolution toward modern hypnotic capacities
and religious propensities.

The *Homo erectus* female in our story also revealed a capacity for hypnotic experience. Her brain allowed innovative ideas to bubble forth from
her unconscious mind, producing creative visions and other unusual experiences. When *Homo sapiens* developed language, healing rituals became
more therapeutic because the rituals were coupled with suggestion. Because
hypnotizability is linked to anomalous experiences such as apparitions, extrasensory perception, and out-of-body experiences, the ritual processes selecting hypnotizability genes also affected the incidence of these episodes.
These anomalous perceptions generated and shaped belief in spirits, souls,
life after death, and magical abilities, beliefs that provided the foundation
for shamanism, the first religious form.

This *ritual healing theory* explains religion's origin and posits that the
processes by which religion began still occur today. The theory is controversial, but it can be evaluated empirically by researchers within the fields
of anthropology, folklore studies, ancient history, neurophysiology, medicine, and the sociology and psychology of religion. *Wondrous Healing* reviews the evidence for the ritual healing theory. The title refers to both spiritual healing and to other forms of anomalous restoration, some of which
are seemingly spontaneous in that they are unplanned. Spiritual healing is
defined as the restoration of physical, mental, emotional, or spiritual
health ostensibly through occult, supernatural, or paranormal means. Examples of spiritual healing methods include faith healing, prayer, shamanism, and various forms of folk healing. The ritual healing theory argues
that wondrous healing has had evolutionary impact, shaping the biological
propensity for religious belief and ritual.

As I observed spiritual healers in Asia and the United States, I came to realize that very different cultures often use similar spiritual healing practices.
I have also discovered that certain types of people benefit from these practices to a greater degree than others. Three examples portray the main features of the ritual healing theory.

BANGKOK, THAILAND, 1985

Wilasinee, an attractive former school teacher, found that she could go
into trance and allow spirits to speak through her.[1] She began healing relatives and neighbors, and when word spread of her success, her spirits devised a healing ceremony that attracted people from all over Thailand.

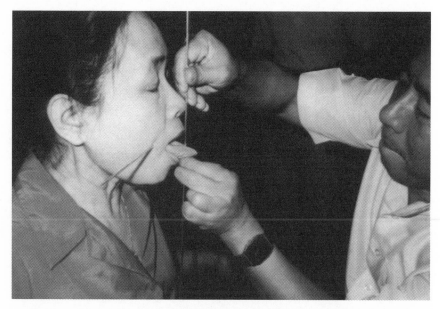

Skewers are inserted in the Thai shamanic healer Wilasinee's cheeks and tongue.
(photo by J. McClenon)

Wilasinee told me that her treatments were most suitable for "people doctors cannot help."

Wilasinee's ceremony included trance performance, symbolic gestures, preaching, and wondrous feats. After she was introduced to the audience, she addressed them briefly, then went into trance. The spirit who spoke through her had a light-hearted attitude yet proclaimed his power to overcome infirmities. Wilasinee extinguished a huge bundle of glowing incense on the palm of her hand, then revealed that her hand was uninjured. The audience gasped in amazement. With the help of assistants, she inserted silver skewers through her cheeks, tongue, arm, and hand. With a huge skewer passing through her tongue, she told jokes, making the audience laugh at her slurred speech and clownish behavior. She flirted with an elderly woman in the front row. "She is possessed by a spirit who loves women," someone explained to me, laughing. The incongruity of the situation captured people's attention. Wilasinee showed no discomfort, and none of her wounds bled.

After the needles were removed, the spirit speaking through her interviewed each supplicant, asking about medical and personal problems. Wilasinee maintained a professional yet concerned demeanor. Sometimes she seemed to know of a person's infirmity without being told. Wilasinee's extended family also performed, waving knives about people's infirmities to cut away their problems. Her brother-in-law rapped rhythmically on a table with a special stone. This caused one man to writhe about on the floor until the demon thought to possess him departed. "He suffers from a mental

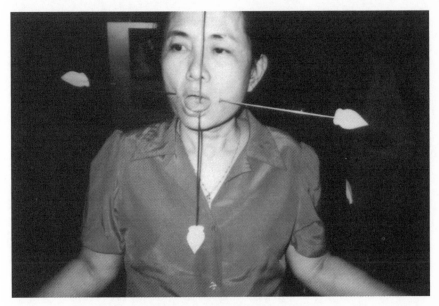

Wilasinee talks with skewers in place. (photo by J. McClenon)

disorder or maybe a devil," someone told me. "This treatment is designed to take care of those things." The high point of the performance was the final healing ceremony. Wilasinee went into trance, and the supplicants were brought before her. She placed her foot on a red hot iron plate and then on each person's afflicted body part. Her foot remained unharmed.

I talked with audience members who agreed that the rituals had been successful. Some described miracles that they had experienced or witnessed. "Last year, I was virtually blind," one woman told me. "After Wilasinee's ceremony, my sight returned completely. Now I come to tell my story and to see the healing of others." Although others told similarly wondrous stories, I noticed that an arthritic woman beside me had not gained greater range of motion in her hands and fingers. Others also appeared not to have been healed, no matter what they claimed.

"Only some people benefit," I thought.

Wilasinee's ceremony is an example of a shamanic performance.[2] Shamanism is a religious system in which the shaman, or spiritual practitioner, goes into trance in order to contact spirits thought to affect living people. Shamanism is the oldest religion, and it is the basis for all later religious forms.[3] Shamans may fly spiritually (mentally traveling to a distant place), allow their bodies to be taken over by spirits (by way of trance possession or mediumship), demonstrate heat immunity (handling or walking over burning coals), exhibit pain denial (piercing or cutting their bodies), and perform spiritual healing or cursing. The most typical forms of shamanic performance involve a practitioner going into trance, communicating with spirits, and gaining remedies or effecting cures for those seeking healing. Although shamanic healing is an archaic religious form, it also occurs in modern societies.[4]

Shamanic performances generally include therapeutic suggestions that are often inferred. We might speculate that the Thai woman who was healed of her blindness suffered from a psychological problem that prevented her brain from processing signals received from her eyes. Wilasinee's ceremony was probably a catalyst for cognitive or even physiological changes that allowed her brain to begin functioning normally. The woman's returned vision was a product of Wilasinee's suggestions and the social atmosphere she constructed. Although Wilasinee believes in the spirits who speak through her, she understands the power of suggestion and belief. "The spirits are the healers, not me," she explains. "The people see that the spirits have power to overcome heat and pain. That brings about belief which results in healing."

People sometimes label methods like Wilasinee's "faith healing." If a person believes that some benefit will be gained from a ritual, the body may heal itself. Wilasinee's ceremony shows that shamanic performances can induce faith. It is clear that she has special talents. Most people cannot go into trance, extinguish burning incense on their palms, puncture their skin without bleeding, or touch a red hot iron grill without blistering. Wilasinee's performance captures people's attention, and they assume that the spirits allowing these things can heal.

Two well-established processes help explain spiritual healing: the placebo effect and hypnosis. These two mechanisms overlap, but differ. Placebos are actions or substances that are devoid of pharmacological effect, but are given for psychological effect. Placebos require belief, and their effectiveness is based on expectation. A person who believes that a specific activity is curative can gain benefits from engaging in it. Placebos can cause release of endorphin—a natural opiate that reduces pain— in the brain. Hypnosis differs from the placebo effect because it depends on a special trait, hypnotic suggestibility, or hypnotizability. Hypnosis is "a psychophysiological condition in which attention is so focused that there occurs a relative reduction of both peripheral awareness and critical analytic mentation, leading to distortions in perception, mood, and memory which in turn produce significant behavioral and biological changes."[5] Hypnosis requires hypnotizability, a trait that allows certain people to respond more fully to ritual suggestions. Hypnotic treatments can be successful even if the person does not believe in them. The hypnotic reduction of pain does not depend on the release of endorphin, but hypnosis can induce belief and expectation, and, as a result, create placebo effects.[6] Both hypnosis and the placebo effects can induce physiological results: they may reduce blistering after exposure to heat, for example, or reduce bleeding from a wound. If the Thai woman's sight disorder was psychologically based, as seems likely, a hypnotist could cure her.

I have observed shamanic performances in Thailand, Sri Lanka, China, Korea, the Philippines, Japan, and the United States. Everywhere the healers appear to use hypnotic processes to induce trance, to demonstrate heat and pain resistance, and to effect cures. I will argue that hypnotic processes provide a basis for many of the effects associated with shamanic performance and healing. People who are more open to hypnotic suggestion are more likely to be healed.

In chapter 4 I review the evidence indicating that hypnotizability can be measured using standardized tests, that it is genetically based, and that it remains relatively stable throughout one's life. People who are highly hypnotizable do not require special suggestions to respond hypnotically, but their response is facilitated by hypnotic inductions.

The ritual healing theory describes how shamanic healing increased the frequency of genes related to hypnotizability. Shamanic healing has probably been used for more than 30,000 years, and its ritual precursors, with their associated physiological bases, probably existed among hominids long before. Hominids who were more hypnotizable had a survival advantage, and their genes related to hypnotizability were passed on to later generations. For example, the Thai woman cured of blindness would be more likely to attract a husband and to rear her children successfully as a result of her cure. She and the others who were healed would be more likely to pass on their genes to future generations than those who were not healed. Over time the frequency of genotypes related to hypnotizability would increase. Even if only 1 percent of a community were cured by shamanic healing, the effectiveness of the healing would have a major impact over thousands of years.

RURAL MINDANAO, PHILIPPINES, 1985

In 1984 and 1985, I witnessed over one thousand psychic surgeries, interviewed dozens of Filipino healers, and talked with perhaps a hundred patients. I watched Josephina Sisson perform psychic surgeries at her chapel in rural Mindanao, Philippines. She went into trance and then seemingly cut patients' skin with her bare hands. She appeared to remove diseased organs and then magically caused the wound to heal without a scar. Sometimes she did a kind of brain cleaning procedure. She caused a piece of cotton to seemingly disappear into the patient's ear and then she apparently removed the cotton from the other ear. By watching closely I have perceived that most "surgeries" involve sleight of hand. The healers conceal pieces of animal organs and small packets of red liquid within their hands; they break open the packets and make it appear as if they have put their fingers into pooled blood. They then deftly pretend to extract tissue from the "wound." After the surgeons remove their hands, an assistant wipes up the "blood," and no scar is revealed.[7]

Psychic surgery in the Philippines is actually an extension of indigenous shamanic performance. All over the world some shamans go into trance and then use sleight of hand to create the illusion that they are extracting

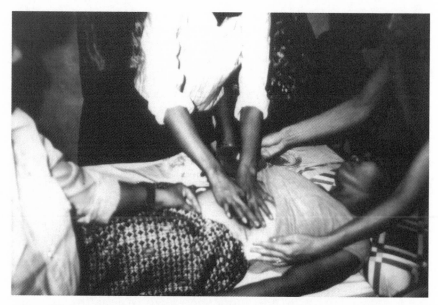

Filipino psychic surgeon seemingly removes tissue with his bare hands.
(photo by J. McClenon)

symbolic objects from their patients. Sometimes they suck a diseased area and then vomit out "impurities." Following World War II, Filipino Christian spiritualists modified this motif to fit Western images of surgical operations. Today foreigners, mostly from the United States and Europe, come to the Philippines in search of cures, and various healers have become wealthy as a result of these visitors' gratitude.

Josephina Sisson lives in a rural area, so most of her clients are Filipino. That is why I was surprised to see an extended family from Pakistan seeking healing. "I have documents proving my claim," a middle-aged Pakistani man told me fervently. "My doctor made x-rays and ran many tests. He told me that I had stomach cancer. Last year I came to the Philippines for a psychic operation. When I returned to Pakistan, my doctor found no cancer. I was cured." I had heard many stories from others who had been restored to health. They all believed that something miraculous had occurred, but described symptoms that could have had psychological bases. At the time I was seeking claims that defied scientific explanation. Because most spiritual healings seemed psychosomatic, I regarded most accounts of them to be insufficiently wondrous to be of interest. Although many people claim benefits from psychic surgery, only a small percentage are fully healed, and virtually all of these complete healings can be attributed to hypnotic or placebo effects. The Pakistani man might have experienced spontaneous remission, or perhaps his disease had been misdiagnosed. Sleight-of-hand magic is hypnotizing; it captures people's attention, allowing the event to implant hypnotic suggestions. Spiritual healings may seem remarkable, but they are more common than many people realize.

"Would you like me to send you the documentation?" the Pakistani man asked.

I considered his offer. I had no way to translate the man's medical records into English. I was seeking accounts that defied scientific explanation, and I doubted that his case would qualify. "No, I don't think so," I replied. "I've heard a number of similar stories. I'm mainly interested in the healers. Many describe special experiences. They tell of having dreams that come true or of receiving messages from people who have died."

"I have had dreams that come true," he told me solemnly. "One dream directed me to come to the Philippines."

At the time, I did not recognize the pattern. People who have many unusual experiences are often involved with spiritual healing. They develop powerful forms of belief that allow them to be healed, and some become healers themselves.

I watched Josephina do "surgeries" on the Pakistani man's wife and children. The wife argued with Josephina, demanding extra surgeries for the children. Josephina stated that further operations were not required. She stated that what was needed was faith in Jesus Christ. The family was puzzled. They were Muslim.

"We wish extra surgeries so we can be sure our children will stay healthy during the coming year," the wife explained to me. Apparently, the Muslim woman regarded the psychic surgeries not as the equivalent of Western operations but as a form of preventive medicine. The more one obtained of this magical treatment, the woman reasoned, the better. Although her children's "surgeries" had no obvious effect, she wanted to get her money's worth.

Josephina explained to the Pakistanis, "At no time did Jesus ever say 'I will heal you.' People are healed by their faith. I do not heal. The surgeries do not heal. Jesus heals, using your faith."

Eventually, both Josephina and the woman departed, frustrated and perplexed. "Faith in Jesus is not required," the woman told me as she left. "My husband is Muslim, yet he was healed."

I agreed with the woman. Apparently belief in a particular ideology was not required. Witnessing Josephina's wondrous performances could trigger sufficient faith to effect a cure.

Spiritual healing reveals similar features all over the world. Healers provide both direct and latent suggestions. Those who capture the imagination of their audiences tend to be more successful. Humans' capacity to react physiologically to suggestion is enhanced when the performance engrosses their full attention—when it is hypnotic. A Paleolithic healer could perform effective therapeutic rituals without using complex religious doctrines or even rapid language.

Another core feature of spiritual healing is its link with spontaneous anomalous experiences. The Pakistani man's dreams led him to seek spiritual healing from a healer whose culture was different from his own. His anomalous experiences, not his Muslim culture, led to his cure. Many people cured by shamanic healers have a similar propensity toward anomalous experiences; research indicates that hypnotizability is correlated with the incidence of other types of unusual experiences. People who are highly hypnotizable tend to report precognitive dreams, waking extrasensory perceptions, apparitional experiences, out-of-body experiences, sleep paralysis, psychokinesis, and other wondrous events.[8] Anomalous experiences provided a foundation for the first religion, shamanism, and they continue to shape folk religious traditions all over the world.

NORTH CAROLINA, UNITED STATES, 1994

Each semester since 1988, my Introduction to Anthropology students interview their friends, relatives, and neighbors, asking, "Have you had a very unusual experience? Would you describe it?" As a result of their work, I have collected over 1,500 narratives of anomalous experience. The purpose of this research is not to prove that supernatural or religious claims are valid, but to determine which types of experience people talk about and the degree to which these experiences are universal.

The narratives indicate that the majority of unusual experiences occur spontaneously, outside of ritual environments. One student collected this story from his grandmother:

> I was involved in an accident that left me partially paralyzed. Because I was in my late forties, the doctors said the bone tissue wouldn't heal properly, so I probably would never walk again. After that, all the days seemed to run into one drab routine because I was not able to do most of the things that I always took for granted. But after several months of what seemed like darkness, the sun finally shone again. I awoke to a familiar voice calling me from the next room. I sat up in my bed and the voice seemed to get louder. It was my late husband calling me from the next room. I sprang out of my bed and ran into the room where the voice was coming from, but by the time I got there it had stopped. So I began searching the entire house but still I came up with nothing. Then it happened: I passed by a mirror and caught the reflection of myself, then once again the voice appeared and said, "See, honey, I knew you could do it!"

Although it is possible that the grandmother's husband actually communicated with her, the student interviewer provided a scientific interpretation of the account: "My grandmother is convinced that anything is possible if you just believe. Maybe that's true, but in my opinion it was her fear that paralyzed her, not the accident. Therefore, in the instant that she heard my grandfather's voice her fears were taken over by a comforting voice."

This story illustrates the connection between wondrous experiences and wondrous healing. People who report anomalous experiences are more likely to experience healings. Hearing an apparitional voice, for example, seems to involve bringing forth information from the unconscious mind to the conscious mind. Although it is possible that the woman heard her husband's actual voice, it is likely that a second person present in the grandmother's house would not have heard anything. For most anomalous experiences to occur, unconscious perceptions must bubble up into consciousness.

In an alternate process, spiritual healing requires reaction to a suggestion. A command that is consciously perceived passes into the unconscious mind and has a physiological effect. For spiritual healing to occur, suggestions must affect physiological processes that are normally controlled outside of consciousness. Some people seem to have thinner cognitive barriers than others. They allow a greater flow of information in both directions across cognitive barriers. I suspect that my student's grandmother had thin cognitive boundaries. This allowed her unconscious mind and conscious mind to interact in a manner that caused her to hear her husband's voice and enabled her to stand and walk.

An alternate hypothesis is that a supernatural force, such as her husband's spirit, caused the woman's perception of his voice and her healing. The supernatural theory cannot be evaluated scientifically, but it does not contradict the previous hypothesis. Both supernatural and natural processes may occur simultaneously. People who perceive apparitions seem to have special cognitive capacities because not everyone has these experiences. It is impossible to scientifically test the argument that people live after they die, but we can evaluate the hypothesis that those who have more anomalous experiences tend to be more hypnotizable. Studies demonstrate that both hypnotizability and the thickness of cognitive boundaries are correlated with propensity for anomalous experiences (see chapter 4).

The common forms of anomalous perception appear all over the world and generate belief in spirits, souls, life after death, and magical abilities. People in northeastern North Carolina report the same anomalous experiences that I learned of in Japan, China, and other areas of the United States: apparitions, extrasensory perceptions, out-of-body experiences, sleep paralysis, psychokinesis, unusual coincidences, and wondrous healings. Those who see a person who is dead, or have a dream that comes true, or feel their consciousness leave their body, tend to devise and accept doctrines that coincide with these perceptions (see chapters 5 and 6).

Spiritual healing provides survival advantages to those whose genotypes are manifested in thin cognitive boundaries and hypnotizability. Shamanism probably developed as a result of the increased frequency of these genotypes. This scenario coincides with what we know of human religious history. Hunter and gatherer societies were the only human social arrangement during the millions of years of the Paleolithic period. All hunter and gatherer societies observed by anthropologists have shamanic religious systems. This suggests that shamanism provided the foundation for all later religions.[9]

Other Theories of the Origin of Religion

No previous theory explaining the origin of religion is widely accepted. Researchers working within the major paradigms continually refine their theories but remain unable to convert their opponents. Differences between these orientations apparently cannot be resolved empirically. The ritual healing theory, on the other hand, is grounded in evolutionary neurophysiology. This orientation makes the theory more amenable to empirical evaluation.

Ironically, some of the earliest theories of religion included arguments that are consistent with the ritual healing theory. In 1651 Thomas Hobbes speculated that the "Natural seed of Religion" consisted of four things: "Opinion of Ghosts, Ignorance of second causes, Devotion toward what men fear, and Taking of Things Casual for Prognostiques."[10] This early secular theory coupled an experiential factor (ghosts) with social-psychological components (ignorance, devotion to that which people fear, misunderstanding of causality). It provided an alternative to existing explanations that required a specific religious faith.

In 1799 Friedrich Schleiermacher portrayed religious belief as an "intuition of the universe," a "feeling" associated with the first instance in perception when "sense and object become one, before both turn back to their original position."[11] Later Rudolph Otto presented an alternative model based on the concept of *numinousness,* a special, irreducible quality that he believed provided the basis for religious experience.[12] The numinous "creature-consciousness" or "creature-feeling" reflects an "overpowering, inexpressible might" that must be experienced to be understood. Schleiermacher and Otto sought not to explain the origin of religious practices, but to protect religion from scientific analysis. Within their paradigms, religion existed because it was based on something real, something scientists could never fully explain. Their arguments were popular with believers who wished their faith to be safe from scientific refutation.

Schleiermacher's and Otto's ideas are of value as scientific theories only to the degree that they can be evaluated empirically. Unfortunately, it is not certain that the forms of mystical experience they emphasize occur frequently enough to have impact on folk religious belief. Studies that quantify the incidence of religious experiences do not directly support Otto's theory.[13] Rarely do people mention an overpowering, inexpressible might, as Otto might suppose, or an "intuition of the universe" as Schleiermacher suggests. Furthermore, these theories seem to have been designed to preclude scientific evaluation.

Edward Tylor, Herbert Spencer, and F. Max Müller theorized that religion began with rational inferences based on individual experiences. Their theories were extensions of ideas that were then prevalent in popular culture—that people believe because of what they perceive. In 1871, Tylor argued that religion originated with interpretations of dreams, visions, apparitions, and loss of consciousness, and that such events provided a basis for belief first in individual souls and eventually in spirits.[14] More advanced assumptions lead to belief that such spirits, thought to be associated with animals and inanimate objects, influence people's lives. The ranking of spirits according to their power led to polytheism and ultimately to belief in a supreme deity.[15]

Herbert Spencer presented a parallel experiential orientation in 1876. He suggested that the idea of ghosts, based on perception, developed into the idea of gods: the ghosts of important ancestors became divinities. He argued that ancestor worship is the root of every religion and that belief in the spirit of the dead was universal.[16] Müller, "the father of comparative religion," presented a similar empiricist position, arguing that "religion, if it is to hold its place as a legitimate element of our consciousness, must, like all other knowledge, begin with sensuous experience."[17]

These early theories coincide with the ritual healing theory in that they propose that experiences shape belief. They fell out of favor because they did not explain the social and emotional nature of religious practice, because elements within the theories (particularly the theories of Tylor) did not coincide with later anthropological findings, and because later psychological and sociological theories seemed more robust.

By the 1930s, there was a general retreat from evolutionary thought in the social sciences,[18] but Robert Bellah provided an exception in 1964. In a classic article, he argued convincingly that "religion as symbol system" evolved from less complex to more complex forms and that "the properties and possibilities of more complex forms differ from those of less complex forms."[19] Yet, like the early experiential theories, Bellah's analysis was grounded within a cultural perspective rather than a biological, evolutionary perspective.

During recent decades, academics developed a culture of disbelief regarding the supernatural.[20] Psychologists and sociologists have insulated themselves from folk accounts and beliefs and therefore have failed to recognize the prevalence and universality of common forms of anomalous experience.[21] They ignore the experiences that earlier scholars found significant.

PSYCHOLOGICAL THEORIES

Stewart Guthrie describes a pervasive wish-fulfillment paradigm, which argues that "religion may be understood as an attempt to allay fears, anxieties, and dissatisfaction."[22] Sigmund Freud's *The Future of an Illusion* launched this psychological orientation.[23] Freud argued that religion arose as a result of human beings' awareness of their helplessness before nature and the persistence of infantile and irrational thinking. Freud theorized that humans desire a powerful, supernatural parent who will defend them from nature's threats. Religion is the "universal obsessional neurosis of humanity; like the obsessional neurosis of children, it arose out of the Oedipus complex, out of the relation to the father."[24] Freud argued that humans' construction of and understanding of the idea of God is based on their perception of their own fathers. The humanlike figure they create has to be superior to earthly humans so that it can overcome the powerful forces of nature.[25] Religion enhances the sense of inner hopefulness and protectedness by performing the threefold task of exorcising "the terrors of nature"; reconciling people to "the cruelty of Fate," particularly as it is shown in death; and compensating humans "for the sufferings and privations which a civilized life in common has imposed on them."[26] Religious beliefs constitute "illusions, fulfillments of the oldest, strongest and most urgent wishes of mankind."[27]

Although Freud's theory of religion is rarely subjected to empirical evaluation, it established a theoretical foundation embellished by later scholars. Some of these theorists reject Freud's mythical thinking about patriarchy, but they accept the doctrine that religion was a natural product of neurosis.

Bronislaw Malinowski, for example, theorized that the need to cope with death is a main source of religious belief. He argued that primitive religion was originally concerned with providing sacramental value to the "crises of human life."[28] Events such as conception, birth, puberty, and death appeared to be brought under human control by religion. Belief in magic created ritual, which in turn developed into religion largely so that people could understand and cope with life's crisis points. Malinowski emphasized that personal attachments caused people to need to deny personal destruction. This need is "determined by culture, by cooperation and by the growth of human sentiment."[29] As in Freud's theory, religious faith is hypothesized to have become prevalent because it has the capacity to reduce fear of death and related anxieties.

Although many elements within psychological theories cannot be evaluated empirically, some features have been shown to be false. Religion does not always alleviate death anxiety, and it sometimes enhances it. It does not always compensate those suffering from deprivation, and it sometimes generates mental disorder. But religion is not always associated with pathology. Some forms of religious and mystical experience are correlated with psychological well-being.[30] Both totemism and the Oedipus complex are culturally specific, not universal as Freud supposed. Many religions, particularly polytheistic ones, do not reflect parent/child relationships. Buddhist doctrines, for example, do not entail parental projection because Buddhism makes no hypothesis regarding the existence of God or gods.[31]

Yet certain features within the psychological theories coincide with the ritual healing theory. The ritual healing theory argues that rituals reducing the effects of stress provide survival advantages. Genes facilitating this protection are passed on to future generations. Nothing within this argument contradicts the general wish-fulfillment position. Stress causes health problems, and wish-fulfillment processes can alleviate the effects of stress. The ritual healing theory extends the wish-fulfillment paradigm by providing an explanation for the effectiveness of repetitive ritual.

SOCIOLOGICAL THEORIES

Émile Durkheim's *Elementary Forms of the Religious Life* established the major paradigm within the sociology of religion. Durkheim believed that religious sentiment originated with the concept of the totem. The totem, generally an animal, a plant, or a natural phenomenon, symbolizes both a deity and the clan. The correspondence of totem and clan indicates that worship of the totem constitutes worship of one's own collectivity. He hypothesized that religion contributes to social cohesion by representing values that the society requires its members to internalize. All societies demarcate the sacred from the profane; the "two sorts of representation" reflect societal consensus on whether an object is good or bad.[32] Durkheim believed that the symbols of totem and taboo were necessary in religion because people do not recognize their level of dependence on society. Within the Durkheimian sociology of religion, society is "all that is required to arouse the sensation of the divine."[33] Religious beliefs and practices create the cohesion necessary for a society to survive.

Many anthropologists and sociologists find Durkheim's ideas useful. Alfred Radcliffe-Brown observed that the Andamanese people have many taboos concerning the types of food that can be consumed during pregnancy. Violation of these taboos by either the mother or the father causes sickness. Radcliffe-Brown theorized that these taboos attach social importance to food in the hard-to-feed Andamanese society, hence they serve an important societal function. By limiting the consumption of scarce food, taboos contributed to the society's survival.[34] As with other prevalent formulations, this theory could be true, but we have no good way of putting it to the test. Although scientists have found that a pregnant woman may develop food aversions that help prevent developmental malformations of the fetus, this evidence does not coincide with Radcliffe-Brown's argument.

Many modern scholars have adopted Durkheim's orientation for explaining religious practices. Religion is thought to bring believers together into a single moral community and to provide unifying rituals that grant longevity to the society. Structural features within societies tend to correspond with elements of those societies' religious practice; this correspondence seems to support Durkheim's position. For example, some scholars regard the dietary restrictions of Hindus, Muslims, and Jews as socially and ecologically functional[35] and posit that practices such as ritual healing and exorcism fulfill social functions.[36]

Yet Durkheim's orientation has serious flaws. Why is religion universal in all societies even when economic, political, and cultural institutions provide alternate means of fulfilling social functions? Why should religion be regarded as functional when it has led to the destruction of entire societies?

Why are religion and morality thought to be always connected when some religions do not distinguish between the sacred and the profane? W. G. Runciman notes that Durkheim's explanation of religious beliefs does not actually explain anything. Why, he asks, is the worship of society more readily explicable than the worship of gods?[37] Durkheim argues that the sacred is distinct from the profane, but also claims that sacred religion is basically the same as secular science. Durkheim presumes that religion contributes to solidarity, but Durkheimians have failed to generate empirical research supporting this theory. As Calvin Redekop notes, "the integrating power" of religion in society is not a social fact, but a largely unsubstantiated social-anthropological belief.[38]

Many scholars assume that Durkheim's model is supported by historical analysis. In actuality, his position is based on his supposition that one "primitive" Australian aboriginal tribe, studied around the turn of the twentieth century, revealed the nature of all ancient conditions. The evidence available to Durkheim was incomplete. Careful analysis of the historical case of Iceland refutes Durkheim's theory. Icelandic religion was powerfully shaped by spiritualist phenomena, which were not part of Iceland's heritage, during the beginning of the twentieth century. This observation refutes the argument that religion always reflects the social collectivity.[39]

Perhaps most difficult to explain within the Durkheimian paradigm is the relationship between the individual and the group.[40] Malinowski argues that Durkheim's theories "make religion something that satisfies needs which are entirely autonomous and have nothing to do with the hardworked reality of human existence."[41] Even the notion of the collective consciousness as the basic religious experience is suspect. As E. E. Evans-Pritchard writes, "No amount of juggling with words like intensity and effervescence can hide the fact that [Durkheim] derives totemic religion . . . from the emotional excitement of individuals brought together in a small crowd, from what is a sort of crowd hysteria."[42] The process Durkheim portrays does not fit what researchers observe. Studies reveal that solitary religious experiences are far more prevalent than group religious experiences.[43]

In addition, Durkheim's model does not mesh with modern biological evolutionary theory. Durkheim might argue that religion has no biological basis because he believes that society is "all that is required to arouse the sensation of the divine."[44] This position is refuted by studies of twins and by other empirical evidence indicating that religiosity has a genetic basis.[45] If Durkheimians were to acknowledge a biological basis for religion, they would have to specify stages within a process of natural selection. Durkheim's theory does not explain how a human group with a small amount of effervescence, crowd hysteria, or religious sentiment has a survival advantage over one that lacks a faint tendency in that direction. Primates demonstrate group cohesiveness without the benefit of religion. We should not assume that primate groups that exhibit some amount of crowd hysteria have a survival advantage over those that do not.

The Durkheimian paradigm specifies that groups benefit from religious rituals because they result in increased cohesion, and it assumes natural selection at the group level. Most anthropologists reject the group-selection argument.[46] The theory of group selection is thwarted by the fact that modern wild primate groups exhibit genetic interchange between groups. There is little reason to think that the processes that selected our distant human ancestors differed radically from the processes at work in other animal species.

On the other hand, Durkheim does make some observations that mesh with the ritual healing theory. Durkheim describes the group effervescence that is associated with ritual. *Effervescence* implies hypnotic processes and altered states of consciousness. Hypnotizable people who perform repetitive rituals become more suggestible and accept whatever ideology supports their ritual practice. Rituals suggesting group cohesiveness trigger such cohesiveness among hypnotizable participants. The sociological and ritual healing theories are not mutually exclusive. But extreme forms of functionalism, those that deny a biological basis for religious sentiment, cannot accept biological scenarios. Durkheim stated that social, rather than biological, factors bring about religious sentiment,[47] but it is more logical to assume that biological, psychological, and sociological factors operate simultaneously.

ANTHROPOMORPHISM AS THE ORIGIN OF RELIGION AND OTHER THEORIES

Guthrie argues that pervasive and universal anthropomorphism is the basis for religion. His theory is innovative in that it manifests an evolutionary perspective. Guthrie argues that early humans found it extremely important to be able to identify other humans: "When we see something as alive or humanlike, we can take precautions . . . [or] try to establish a social relationship. If it turns out not to be alive or humanlike, we usually lose little by having thought it was. . . . We see apparent people everywhere because it is vital to see actual people wherever they may be. . . . Faced by uncertainty, we bet on the most significant possibility"[48] by assuming that unidentified, ambiguous objects are living.

Guthrie contends that human perception of humanlike qualities in nature is the first stage in the development of religious ideologies. "Religion consists of seeing the world as humanlike and arises because doing so is a good bet even though, like other bets, it may fail."[49] Because successful quick identification of other humans provides survival advantages, genotypes supporting this behavior increase in the human population.

According to this theory, humans devise anthropomorphic religious ideologies because they tend to label unexplained events as events that pertain to living beings. Guthrie hypothesizes that anthropomorphism is synonymous with religiosity because "anthropomorphism is the core of religious experience," which "springs from a powerful strategy" that "pervades human thought and action."[50] "Religion not only anthropomorphized (as everyone admits) but *is* anthropomorphism."[51] Some religious systems, such as Buddhism, are not based on anthropomorphism, but Guthrie argues that "these systems are ethical, philosophical or psychological, not religious."[52]

Although Guthrie's theory seems logical to some, it is difficult to devise studies that could test it. Individuals who are quick to engage in anthropomorphism have no observable survival advantages, and people who often see living beings where there are none are at a survival disadvantage. Furthermore, a large percentage of the unusual experiences people report do not seem to be related to anthropomorphism, as might be hypothesized by Guthrie. People tend to attribute out-of-body experiences, extrasensory perceptions, and various other forms of unusual perceptions to cosmic structures rather than to living beings.[53] Also, Guthrie's theory would suggest that those who have the greatest tendency to anthropomorphize would be the most religious. No studies indicate that this is the case.

On the other hand, many forms of unusual experience, such as psychokinesis and sightings of apparitions, can be interpreted as reflecting anthropomorphic processes. For example, the woman who was able to walk after she heard her deceased husband's voice may have heard her husband's spirit (the folkloric explanation), heard a mental creation generated through anthropomorphic processes (the anthropomorphic explanation), or heard a mental creation that was *not* generated through anthropomorphic processes (an alternate explanation). We have no way to determine which possibility is valid.

Guthrie's anthropomorphic explanation meshes with the ritual healing theory in some respects. According to the ritual healing theory, a highly hypnotizable person participating in a ritual would be more likely to accept whatever suggestion was provided. Anthropomorphic explanations seem logical to most people in many cases. Therefore, a highly hypnotizable person would be likely to accept an anthropomorphic explanation. Although Guthrie's argument that religion is anthropomorphism (and nothing else) remains unsupported, a limited version of the anthropomorphism theory is consistent with the ritual healing theory.

Various theorists, such as Michael Winkelman, John Schumaker, Eugene d'Aquili, and Charles Laughlin, provide evolutionary models that link cognitive processes to shamanic and religious experience.[54] Chapter 4 reviews some of the evidence supporting their arguments, and those wishing a more complete discussion should consult their work. These theories fall within the same evolutionary paradigm and tend to support each other. Evolution is a complex process, and the ritual healing theory should not be regarded as the only valid explanation for the origin of religion.

A General Critique of the Theories

In Guthrie's summary of the scholarly discourse regarding theories of religion he notes that social scientific theories are diverse and that they share little more than the claim that gods do not exist and that religion is a human creation.

Indeed, humanistic theories of religion are in disarray. E. E. Evans-Pritchard wrote twenty-five years ago that "either singly or taken together, [they do not] give us much more than common-sense guesses, which for the most part miss the mark." Clifford Geertz wrote shortly afterward that the anthropology of religion was in a "general state of stagnation" and lacked any "theoretical framework [for] an analytic account." Most writers still agree. Murray Wax sees continuing "theoretical stagnation." J. S. Preus finds "evidence of an identity crisis" in incommensurate modern approaches to religion. In short, the consensus is that there is no consensus, and little optimism.[55]

Yet the prevalent theories seem to capture pieces of the puzzle. Religion is rooted in emotion, passion, and feeling (Otto, Schleiermacher), yet it has rational components that are derived from experience (Müller, Spencer, Tylor, Guthrie). It has emerged from the unconscious (Freud), yet it has a foundation in social experience (Durkheim). All of the prevalent theories regarding the origin of religion seem to be partly true, but we have no way to determine whether alternate explanations are also true.

These prevalent theories are like religions themselves. They are popular because people wish to believe in them. They allow scholars to explain away religion and do not provide predictive ability or practical applications. The evolutionary paradigm and the ritual healing theory tend to absorb these previous paradigms. The ritual healing theory explains religion's emotion, passion, and feeling; its rational components that are derived from experience; its emergence from the unconscious; and its foundation within social experience. Hypnotic group rituals associated with emotion, passion, and feeling facilitate anomalous experiences to emerge from the unconscious. Both rational and emotional evaluations of these perceptions contribute to religious belief.

Unlike previous orientations, the evolutionary paradigm is subject to empirical evaluation within the fields of animal behavior, physiology, medicine, history, anthropology, sociology, psychology, and folklore studies. Whereas prevalent social science paradigms emphasize cultural factors that affect religion, the ritual healing theory posits that religion has universal features with physiological bases. The empirical foundation for this theory ensures that through research we can gain ever greater insights into the evolutionary processes that shape religious practice.

Daniel Dennett writes that Charles Darwin's idea is like a "universal acid" so powerful that it eats through whatever vessel attempts to contain it.[56] Some social scientists may seek to thwart evolutionary encroachments into their fields (see the concluding chapter of this volume), but their efforts will inevitably be unsuccessful. Discoveries within physiology, genetics, evolutionary social psychology, and the study of animal behavior continually carve away at academic domains thought safe from Darwinian intrusion. In the end, evolutionary social sciences will emerge, and Darwinian theories explaining the origin of religion will gain acceptance.

1

The Evolution of Wondrous Healing

"I'm asking you to simulate a pre-linguistic hominid campfire ritual," I told the people attending my presentation at an academic conference. "Please move to the front of the room and sit in two rows, each row facing the other." As about twenty people moved, one man refused to participate. I explained that refusals occur in all groups, that few groups are unanimous, and that rituals do not inevitably lead to social cohesiveness.

"I ask that you pretend that you have no language. You can make only a limited number of sounds. The men will call out 'Hey!' and the women will call out 'Who!'" I then led the group in the exercise, having women and men call out alternately—the males calling "Hey, hey, hey," and the women responding, "Who, who, who." They fell into a lengthy chant. It was like a song. Some people enjoyed the ritual more than others. Such chants induce altered states of consciousness in some people. One woman smiled and allowed her eyelids to fall closed; another scowled and shifted uncomfortably in her chair.

"When hominids gained limited verbal ability, their chanting evolved into singing and eventually into language," I explained. "Our hominid ancestors selected each other in part based on vocal ability. Let's include another element in our exercise. Our ancestors were all successful at mating. I ask that you continue chanting, but use your voice and eyes to try to attract members of the opposite sex." The smiling woman continued to enjoy the exercise and flirted with many men. Some people looked deeply into each other's eyes. Others participated half-heartedly. The frowning woman became more hostile.

I brought the chanting to an end and asked the smiling woman if she would volunteer for a hypnosis demonstration. I explained that her behavior suggested that she would be a good hypnotic subject because the chanting had caused her to become more relaxed. She agreed to participate, and I suggested that she close her eyes and imagine that her right hand was becoming lighter and lighter. Almost immediately, her hand started to rise, progressing in the small, jerky movements characteristic of a hypnotic process. I suggested that her hand would move toward her face and that when it touched her face, she would fall into a deep trance. Within a minute, her hand touched her face, her head fell forward, and she appeared completely entranced.

I explained that the chanting exercise has hypnotic qualities and that some people are more hypnotizable than others. Pre-linguistic hominids undoubtedly discovered this phenomenon and used chanting and drumming for amusement and relaxation, as do tribal people today. Language was not required. I argued that the smiling woman might have become a shaman in a hunter and gatherer society because of her hypnotic ability. I suggested that she could leave her body and travel to heaven, where she could visit any departed relative that she might wish to see. She described drifting upward into the clouds, meeting her deceased grandmother, and carrying on a conversation. I suggested that she bid her grandmother goodbye, go back to her body, and return to a normal state of consciousness. I then suggested that she open her eyes and feel relaxed and refreshed. I had various participants ask her about her experiences. She described her images as "very real . . . perhaps more real than normal life." Although people's reactions to my exercise varied, some religious people felt that the woman had actually contacted a spirit. Many people were startled and entertained by this simulated shamanic ceremony.

"This exercise illustrates how an ancient repetitive ritual would have had evolutionary impact." I explained. "Some people are more hypnotizable than others, and these people gain greater benefits from ritual activities. Our ancestors not only coped with predators but they also engaged in rituals as a form of recreation. Ritual behavior shaped their music, their mating, their art, their hypnotizability, and eventually their religion. Because those who gained the greatest benefits from these rituals had survival advantages, repetitive rituals selected for genes associated with hypnotizability. The process led some people to perform shamanic roles."

"I'm very skeptical," the angry woman exclaimed. "I don't believe in hypnosis. It is merely role-playing and has no basis in fact. Anyone can play a role."

"Your argument has merit," I noted. "People are continually playing roles, and the hypnotic and shamanic roles are some of them. But if we used an electroencephalograph, or EEG, to measure each person's brain waves during the chanting exercise, we would find that some people's brain-wave patterns were affected more than others. Certain people whose brain-wave patterns were more affected have a physiological propensity for hypnotic and shamanic experience. I have found that when I use good hypnotic subjects, I can induce a wide variety of religious experiences: hallucinations, spirit communication, and other trance behavior. This exercise portrays how repetitive rituals lead some people to experience trance and to engage in performances supporting shamanic belief systems."

The angry woman gave no response. It is difficult for people who have little hypnotic propensity to understand the sensations gained by those who do. The smiling woman talked with me after the session. "I really enjoyed the ritual," she stated. "But I suspect that if some of these academics lived during pre-linguistic times, they wouldn't get mated," she joked. Although it is doubtful that the hypnotically challenged had severely reduced fertility, it would seem that they would be at a disadvantage living in an pre-linguistic environment where the main form of recreation was chanting or singing around a fire.

My exercise illustrates one element within the ritual healing scenario: the evolutionary progression from animal to human religious ritual was probably associated with hypnotic processes. The anthropological and historical evidence that is reviewed in this chapter allows a clearer portrayal of this scenario. Ancient primates' physical and social environments led to rudimentary hypnotic capacities. These primates used simple rituals to communicate and to alleviate stress, and hominids devised more complex rituals. Middle Paleolithic *Homo sapiens* devised therapeutic rituals based on the use of symbols. During the Upper Paleolithic period, *Homo sapiens* left cave paintings and musical instruments, suggesting complex ritual symbolization associated with altered states of consciousness. At some stage, *Homo sapiens* devised shamanic healing systems based on altered states of consciousness and communication with spirits. By the time humans began writing, all known medical systems were intertwined with religion. We can evaluate this scenario in light of animal studies, anthropological evidence, and historical analysis of ancient medical practices.

Animal Studies

The evolution of species-specific behavior involves an interplay of genetic and environmental factors. An animal's developmental process follows prescribed stages, and an organism is more susceptible to certain external influences during specific time periods. An animal of one species learns a behavior at a particular age more easily than an animal of another species does. Birds, for example, have an innate tendency to sing in a prescribed manner, but they learn "proper" songs by listening and practicing when they are young. Songs vary within a species depending on geographical location. Animals living in groups reveal rudimentary elements of what is termed culture in humans.

Many species demonstrate special types of startle responses that appear to be simple forms of hypnotic behavior.[1] Such reactions, shaped through evolution, are valuable for the species survival. Animal catalepsies, or pretended deaths, seem to be derived from the animals' instinctive knowledge that predators generally do not attack the dead except in cases of extreme hunger. When the motionless animal feels safe, the catalepsy disappears. This Totstell reflect has been labeled animal hypnosis because many of its manifestations appear similar to those of human hypnosis. The animal seems to experience a rapid transformation of consciousness that affects its behavior.

Although it is not certain that the human hypnotic capacity evolved directly from ancient versions of the Totstell reflect, a connection seems likely because both animal startle responses and human hypnosis involve similar environmental cues, changes in consciousness, and behavioral responses. Common methods for inducing an animal's startle response (application of repeated stimuli, application of pressure to the body, restriction of mobility, and sudden changing of position) are cues that can also induce human hypnosis. Animals that demonstrate startle responses react to inductive stimuli with a variety of behaviors. They may hide between stones or exhibit the dreamy, stuporous behavior that is observed in humans. The reactions appear to have been shaped by evolutionary processes. Both animal and human reactions involve a sudden shift in consciousness toward passivity and paralysis; these reactions differ from flight-or-fight reactions.[2]

J. Hoskovec and D. Svorad documented a number of features regarding hypnosis in rats. First, susceptibility to animal hypnosis decreases with individual development. Second, there are individual differences in susceptibility within a species. The response can be evoked in about 20 percent of rats. Third, animals that are susceptible to animal hypnosis share common characteristics. For example, the more susceptible rats demonstrate a higher degree of locomotive activity and a lower degree of nonlocomotive activity (such as scratching, biting, washing, and yawning) than other rats. Fourth, there are metabolic differences between animals susceptible to animal hypnosis and those that are not susceptible. Fifth, electroencephalographic findings regarding animal hypnosis vary.[3] Hoskovec and Svorad's studies suggest that the capacity for trance behavior exists in rudimentary form in rats and that evolutionary selection of hypnosis genotypes has affected a variety of peripheral traits.

Within some species, particularly those that live in groups, hypnotic processes are linked to ritual behavior. Repetitive stimulation that induces animal hypnosis is parallel to the repetitious qualities within human rituals. Mesmerists who experimented during the mid-nineteenth century, before the word *hypnosis* was coined, found that repetitive motions were effective for mesmerizing animals, and many people called the force that made mesmerism possible "animal magnetism."[4] Modern researchers note similar equivalencies between chimpanzee and human responses to repetitious stimulation.[5]

Ritual behavior is particularly prevalent among species for which cooperative behavior is important.[6] Animal observers refer to *display* activities and *social releasers,* or instinctual activities that have been displaced from their usual functions and exaggerated or modified to stimulate particular reactions in other members of the same species. It is generally thought that rituals aid organisms in reducing tension because the repetitive, choreographed movements are associated with relaxation and fixation of attention—features associated with hypnotic induction. In many species, ritual behaviors reduce anxiety before the animals achieve the close proximity required for mating. The repetitive characteristics of both animal and human rituals seem designed to induce therapeutic alterations of consciousness, and repetitive movements may have therapeutic value even when the animal or person is in isolation. Animals in captivity sometimes adopt monotonously repetitive motions, apparently as a means of reducing stress.

Human hypnosis appears to have been shaped by the rituals that ancient primates devised to cope with group life. Chimpanzees, rhesus monkeys, stump-tailed monkeys, and bonobos use rituals to reduce aggression and facilitate reconciliation.[7] Human physiological response appears to be similar to that of other primates. The same repetitive movements and sounds that hypnotize humans are also effective for hypnotizing chimpanzees.[8]

But the rituals of humans are far more complex than those of other animals. Human rituals are performances of more or less invariant sequences of formal acts and utterances that are not encoded, or created, by the performers.[9] Eugene d'Aquili argues that ritual is "a sequence of behavior which is structured or patterned" and "rhythmic and repetitive (to some degree at least)." This allows ritual to "synchronize affective, perceptual-cognitive, and motor processes within the central nervous system of individual participants."[10] Rituals must be performed, but there is room for imperfections and variations in the performance, an aspect that holds audience attention while tending to trigger dissociative states. D'Aquili emphasizes that rituals have physiological benefits bringing about evolutionary change.

Various animal activities suggest that there may be an evolutionary bridge between animal ritual and human religious behavior and sentiment. J. Malan witnessed members of a species of wild baboon (*Cynocephalus porcarius*) that made special efforts to observe the rising and setting sun and remained "in a sort of rapt watching attitude until the sun had gone."[11] Similar solar rituals have been observed in other species, including many of the larger birds and the African meerkat. Some researchers suggest that this behavior illustrates the tendency of many organisms, including humans, to respond sympathetically to light.[12] Researchers have recently noted that light is important for organisms and that images of light are pervasive within mystical experience.[13]

Wolfgang Köhler observed unusual chimpanzee rituals at the Anthropoid Station in Tenerife, Canary Islands.[14] A group of chimpanzees engaged in a "spinning game," or "motion-pattern," reminiscent of the dancing of Sufi whirling dervishes. The spinning dance was associated with rhythmic sounds and sometimes involved self-adornment with a rope, a rag, grass, a twig, or some other object. Köhler saw chimpanzees

> march in an orderly fashion and in single file round and round the post. . . . They no longer walk, they trot, and as a rule with special emphasis on one foot, while the other steps lightly; thus a rough approximate rhythm develops, and they tend to "keep time" with one another. They wag their heads in time to the steps of their "dance" and appear full of eager enjoyment of their primitive game. Variations were invented time and again; now and then an ape went backwards, snapping drolly at the one behind him. . . . I first took part in this game after it had taken place without me hundreds of times. It seems to me extraordinary that there should arise quite spontaneously, among chimpanzees, anything that so strongly suggests the primitive dancing of some primitive tribes.[15]

Jane Goodall observed a kind of rain dance among the wild chimpanzees of Gombe, Tanzania. The animals repeatedly charged down a hillside during a thunderstorm, engaging in threat behavior.[16] Stewart Guthrie considers this behavior a rudimentary form of anthropomorphism.[17] Dancing behavior has been interpreted as a precursor to symbolization that may reflect primate use of therapeutic altered states of consciousness. Goodall did not observe wild chimpanzees engaging in prolonged rhythmic drumming, but she does note that they sometimes create rhythmic sounds when they travel in groups, by pounding on trees with their feet. "Frequently, when a number of chimpanzees traveling together come upon some favored 'drumming tree' along the track (a tree with wide buttresses) each male in turn drums in this manner. This results in a whole series of one to three double beats with irregular intervals between each."[18]

Goodall also describes ritual behaviors that are common among chimpanzees at Gombe.[19] Of particular importance is grooming, a ritual that occurs in a variety of situations but particularly when the animals were tense, anxious, or fearful. Subordinates that approach superiors often reach out and groom them and are acknowledged peacefully. Grooming also is a means for the dominant animal to calm a subordinate seeking reassurance. The animals sometimes engage in prolonged sessions of grooming during which both participants gradually relax.[20]

The grooming response indicates a rudimentary capacity for dissociation and trance. "Sometimes the groomer lazily runs a forefinger through the other's hair, his own eyes drooping."[21] Goodall also notes a form of "symbolic" grooming. A chimpanzee may begin grooming a leaf, engaging in "a displacement activity, not yet properly understood," which attracts the attention of proximate animals who are transfixed by the behavior.[22] Clear parallels exist between leaf grooming, human hypnosis, and shamanic rituals: symbolic performances capture observers' attention and produce psychophysiological effects.

Chimpanzees, like humans, are distracted by rhythmic stroking and can experience pain relief as a result. Animal grooming appears to fulfill a function that is similar to that of dissociative, analgesic hypnosis. Goodall reports:

> When Mandy's three-month-old infant appeared with a torn and bleeding arm from which the bone protruded, her pain was obvious; she held her head rigid with eyes open, glazed and staring. But as Mandy groomed her she temporarily relaxed, and her eyes closed. Once the infant Flint was bitten by an enormous ant, which remained clinging to his brow; he whimpered and rubbed his face against Flo's breast. She was apparently unable to detect the cause of his distress, but groomed him from time to time. Each time she did so he quieted, only to whimper and rub his face again when she stopped.[23]

Chimpanzees not only comfort one another, they also perform extremely simple medical procedures. Goodall notes that chimpanzees have removed splinters from each other's hands and feet and, in captivity, have provided rudimentary treatment of a wound. Beyond these actions, they provide few other services.[24]

> On one or two occasions, one chimpanzee licked the wound of another. On the other hand, when one female had a really bad puncture wound in her shoulder, which she repeatedly presented for grooming to an adult male, he appeared fearful, whimpered, and hurried away from her. . . . We have seen chimps, especially youngsters, staring intently at a wound of another chimp, putting their faces close to it, but not doing anything else, even in the case of a mother carrying a badly wounded infant.[25]

Köhler describes rudimentary treatments among captive chimpanzees at Tenerife:

> Once an enormous abscess had appeared on the lower jaw of one of our chimpanzees. When it became noticeable through the extent of the inflamed surface and secretion of pus, another of the apes would not stir from the patient's side, but pressed and kneaded the injured jaw, until the pus was removed, revealing a raw, gaping wound. The animal thus treated made no objection. . . . The wound—itself probably originally caused by skin treatment with filthy hands—healed rapidly and completely. Chimpanzees also like very much to remove splinters from each other's hands or feet, by the method in use among the ordinary human laity. Two finger-nails are pressed down on either side and the splinter levered upwards, to be caught and removed by the teeth.[26]

Like the health of many animals, chimpanzee health is affected by mind/body relationships. Because basic processes that allow mind/body relationships exist even in single cell organisms,[27] we would expect mind/body interactions to have been shaped by evolution. Goodall notes that the death of chimpanzee Flint's mother when he was eight and a half years old appeared to affect his immune response. Immediately after his mother's death he showed increasing signs of lethargy for six days, after-which he was not seen for four days. Flint's physical condition deteriorated markedly for the last two weeks of his life. Postmortem examination indicated that he had gastroenteritis and peritonitis. "It seems likely that the psychological and physiological disturbances associated with loss made him more vulnerable to disease."[28] Perhaps Flint would have recovered from his grief if chimpanzees had effective mourning rituals and if Flint had sufficient capacity to benefit from such rites.

Goodall has catalogued many ritual behaviors that reduce stress.[29] In response to frustrating situations, chimpanzees engage in charging, slapping, stamping, dragging, throwing, and drumming behaviors. Individuals "seemed to be more relaxed after performing one or more of these displays."[30] Some series of evolutionary stages must have occurred in which ancient primate rituals evolved into the ritual behaviors found among *Homo sapiens*. Human rhythmic drumming and dancing produces measurable altered states of consciousness that are associated with trance and relaxation.[31] It is reasonable to suppose that similar therapeutic altered states of consciousness involving relaxation occur among primates.

Among humans, the *relaxation response* is a specific set of physiological changes (decreases in oxygen consumption, carbon dioxide elimination, respiratory rate, heart rate, blood pressure, and arterial blood lactate) that are associated with decreased activity of the sympathetic nervous system.[32] Procedures for eliciting the relaxation response are similar to those for inducing hypnosis, and the physiological states produced by hypnotic induction are similar to the physiological changes of the relaxation response.[33] The degree to which a person benefits from the relaxation response is correlated with the degree of the person's hypnotizability.[34]

Ritual effects may be related to social position because social rank in animals affects their capacity to relax after experiencing stress.[35] Blood pressure and cardiovascular reactivity of rats under stress are correlated with their position in the social hierarchy.[36] When given amphetamines, monkeys respond in a manner reflecting their social position. Dominant monkeys become more threatening, making more frequent dominant displays, while submissive monkeys become more passive. Change in social position is associated with modifications of amphetamine response.[37] Male baboons' hormonal reaction to stress varies with social status; dominant males release high levels of cortisol in response to stress, but their levels decline rapidly; their resting levels of cortisol are lower than those of subordinate males.[38] We would expect that animals that have greater capacity to dissociate or relax during reconciliation rituals would remain healthier. Because genetic factors influence animal temperament, sociability, and independence,[39] we should expect parallel factors to influence psychological traits such as animal hypnotizability.

Models of hominid skulls (from left to right): *Australopithicus africanus, Homo habilis, Homo erectus, Homo sapiens neanderthalensis* (Neanderthals), and *Homo sapiens sapiens.* (photo by J. McClenon)

Anthropological Evidence

The human invention of language made it possible for *Homo sapiens* to link altered states of consciousness with symbolization. This allowed humans greater control over hypnotic processes. A review of what we presently know about the evolution of *Homo sapiens* allows us to estimate the time available for the selection of genotypes related to hypnotizability and religious sentiment.

Evolutionists point out that the physiology of the brain has an "added on" characteristic. The core Reptilian area of the brain is covered by the later developing Paleomammalian area, which is surrounded by the most recently developing Neomammalian area.[40] The components are not designed for overall efficiency, but are the result of a progression of evolutionary adaptations.[41] The human brain became increasingly asymmetrical as a result of this series of additions.

> Between us and the australopithecine, which walked upright but had an ape-sized brain, stand a few million years: 100,000, maybe 200,000 generations. That may not sound like much. But it has taken only around 5,000 generations to turn a wolf into a chihuahua—and, at the same time, along a separate line, into a Saint Bernard. Of course, dogs evolved by artificial, not natural,

selection. But as Darwin stressed, the two are essentially the same. . . . And in both cases, if the "selective pressure" is strong enough—if genes are weeded out fast enough—evolution can proceed briskly.[42]

Each hominid adaptation fixed itself in the population, changed the social environment, and invited more adaptation, so human evolution was, in a way, artificial. Hominid culture affected the evolutionary process. The ability of social primates to walk upright and to use their hands to fashion and operate tools led to brain specialization. Philip Lieberman notes that a form of grammar is associated with complex motor activities.[43] Rules exist for tasks, and those rules must be executed in sequence. The rule schema is automated in the primate's cortex. The conscious part of the brain is not aware of all of the commands that are required to perform complex tasks. Humans can perform actions without conscious thought; the process involves dissociation, the splitting off of mental processes from the main body of consciousness. This splitting off generates cognitive boundaries, a feature that governs human hypnotic processes.

Experts disagree regarding the time frame of language development among humans. Proponents of "early development" posit a drawn-out series of stages. Skulls of *Homo habilis* (1.9 million B.P.) and *Homo erectus* (750,000 B.P.) reveal the presence of Broca's area, the section of the brain often associated with linguistic ability. Alan Walker, a proponent of the "late development" school, argues that these findings do not verify the early development of rapid language because Broca's area is not always linked with speech.[44]

Determination of when humans gained the ability for language would help us to specify the era during which the first shamanic rituals could have taken place. Shamanism requires individuals to go into trance and to provide benefits to others by telling about their experiences. The ritual healing theory hypothesizes that people developed shamanic ideologies in part by describing their anomalous experiences to each other. These processes require the use of language, but it is likely that pre-shamanic therapeutic rituals occurred during pre-linguistic phases.

Lieberman believes that the early biological substrates for human speech probably developed in the last 500,000 to 250,000 years.[45] The evolution of lateralized neural mechanisms, which are associated with human speech, may have originated with the division of labor between the hands: one hand holds an object while the other performs a task upon it.[46] Chimpanzees demonstrate a weak form of hand specialization. Goodall notes that chimpanzees make tools like "termite sticks" and demonstrate throwing behaviors, tasks that encourage hand specialization.[47]

Charles Darwin suggested that hominids acquired the capacity for song (voluntary modulations of vocal cries) as a stage in the evolutionary process leading to the development of language.[48] Merlin Donald argues that the capacity for song entailed the existence of a nonlinguistic, mimetic culture based on conscious, self-initiated, representational acts.[49] Mimetic behavior includes pointing, making gestures, playing simple games, ritual dancing, chanting, and nonlinguistic singing.

Much evidence supports this scenario. Hominid fossils reveal a gradual cranial enlargement, progressive changes of the basal surface of the skull, morphological asymmetry of the hemispheres, and other markers probably associated with the gradual development of the modern vocal apparatus.[50] Virtually all studies of the structure of intelligence indicate that musical talent is a separate factor from verbal skill. Language acquisition follows a different developmental path than does musical ability. All normal children in all cultures learn to speak around the ages of two or three, but musical ability develops later. In addition, damage to the left temporal region, associated with loss of phonetic control, can occur independently of damage to the right temporal region, which can cause the victim to lose control of vocal volume, pitch, tone, and emphasis. This evidence suggests that hominids or humans acquired singing ability before they acquired language. Mimetic song provides a bridge to speech but affords little additional communicative power.[51]

Mario Vaneechoutte and John Skoyles argue that humans do not have a genetically based "language acquiring devise" (LAD) based on an innate universal grammar, but instead have evolved a "music acquisition devise" (MAD) that allows language.[52] They point out that birds have a similar genetically based MAD that allows some to mimic human speech. The development of a MAD that is shaped by sexual selection and group cohesion is more easily explained according to evolutionary processes than the existence of a LAD.

Mimetic rituals such as chanting, singing, drumming, dancing, and other repetitive behaviors produce altered states of consciousness (ASC),[53] and rhythmic drumming has been demonstrated to affect brain patterns, inducing ASC.[54] The rituals and the rhythms may reduce tension associated with social conflict. Those who achieved deeper states of relaxation as a result of ritual would have had survival advantages over those who remained tense. *Homo erectus* acquired the ability to control fire over 500,000 years ago. It is easy to imagine hominids sitting around a fire and engaging in group chanting before the development of rapid language. Such ritual activities may have resulted in selection for hypnotizability. The simulated ritual that I described in the introduction to this chapter illustrates how pre-linguistic rituals may have produced hypnotic states that eventually facilitated shamanic experiences.

Lieberman suggests that Neanderthals lacked the capacity for rapid human speech but had general cognitive abilities somewhat similar to those of anatomically modern humans.[55] "Neanderthal Hominids . . . may represent a hominid line that specialized for chewing and muscular strength at the expense of speech communication."[56] Although Neanderthals may have been capable of some forms of speech, the near absence of symbolic objects in the Mousterian (130,000–40,000 B.P.) suggests to some anthropologists that language as we recognize it was absent among them.[57]

Yet even if their linguistic ability was limited, Neanderthals left behind relics indicating that they made fires, used tools, and cared for their elderly and infirm. They ritually buried their dead, placing some in burial pits with flower offerings.[58] They left behind circles of goat horns and evidence that they worshipped or honored these animals. Remains suggest rudimentary forms of religious ideology.

> The most conspicuous evidence for definite ideologies in Neanderthal man consists in the burial of the skull No. 1 of Monte Circeo. . . . The Monte Circeo skull, representing a late or typical Neanderthal of La Chapelle-aux-Saints or Neanderthal form, about the age of forty-five at death, was lying on the floor of a cave surrounded by a circle of stones.[59]

The Monte Circeo skull was mutilated, and Alberto Blanc argues that this mutilation was associated with removal of the brain, probably for the purpose of cannibalism. The skull is one of many similar skulls that fit the same pattern. "The mutilation of the base of the skull has been performed by 'early' and 'late' Neanderthals for a very long time, estimated to be about 250,000 years."[60] But if consumers were merely trying to get at the brain, they would have simply smashed the skulls. The careful enlarging of the cranial opening suggests ritual behavior. Although there are alternate explanations for the existence of such relics, it is possible that symbolization began among *Homo sapiens* as early as 250,000 years ago.

Yuri Smirnov's analysis of "intentional human burials" during the Middle Paleolithic period (about 70,000 B.P.) supports the argument that hominids engaged in burial rituals.[61] Based on an evaluation of about sixty cases from eighteen cites (eleven European and seven Asian), Smirnov concluded that classic Neanderthals, Neandertaloids, and the *sapiens* form all began the practice of intentional burial at about the same time (68,000–78,000 B.P.).

> Middle Paleolithic people invented a "cult of the dead," which seems to have been based on pairs of opposites, or dualities, apparent in even the earliest burials: concealment or exposure of the body, burial of the body intact or disarticulated, burial of the whole body or only part of it, burial of the body in an extended or a flexed position (lying on the back or on the stomach, or on the right or on the left side, with the lower limbs extended and the upper limbs flexed or with one limb flexed and another extended, etc.), burial in a depression or under an elevation, burial with grave goods or without, burial with associated features or without, and so on. All these traits can be observed within one period, one population, one site, or one burial group, sometimes, even within one grave. . . . The dualities expressed in the mortuary practices of Middle Paleolithic groups lead us to conclude that they already had the dualistic perception of the world which later became the foundation of almost all culture.[62]

Yet Neanderthal sites contain no convincing evidence of painting, sculpture, engraving, jewelry making, long-distance trade, or the use of formal bone tools, and there is no marked geographical or temporal variation in stone tools. These behaviors are associated with anatomically modern humans who lived during the Upper Paleolithic period. The later-language-development theorists argue that some form of depiction, or art, was necessary as a first step toward the development of a rapid communication system. The earliest relics displaying such symbolization date from 35,000–29,000 B.P. and are associated with *Homo sapiens sapiens.*[63]

The evidence that suggests that Neanderthals lacked rapid language ability but engaged in religious rituals coincides with the ritual healing theory. The practice of religious rituals during early stages of language development implies biological linkage of religious sentiment, ritual activity, and anomalous experiences supporting religious ideology. This link could have been forged before development of rapid language.

At some stage in human evolution, ritual behavior became associated with medical care. Relics dated as early as 50,000 B.P. seem to show that Neanderthals cared for the infirm and injured. A middle-aged man whose body was at La Chapelle-aux-Saints had serious arthritis and had lost most of his teeth before death. He would have required substantial care by the members of his group.[64] A Mousterian skeleton at Shanidar Cave, Iraq, had a crushing bone fracture around the left eye that probably caused blindness in that eye. This was coupled with severe injury to the right side. The right arm had ceased to grow following the injury, and it is possible that the head injury caused nerve damage that rendered the right side weak or immobile. The right foot was also broken. All these injuries had healed long before death. This man would have been dependent on others for many of his physical needs.[65] Neanderthals had the capacity both to care for their injured and to engage in ritual activity, and it seems likely they combined these activities. Therapeutic rituals would have provided greater benefits to those who were more suggestible.

No matter which time frame is appropriate for explaining the development of linguistic ability, scholars agree that a co-evolution occurred, joining human biology and culture.[66] Language and cultural development were inseparable from physical evolution. The biological development of the tongue, mouth, and throat and of the cognitive capacities required for language coincided with cultural advances that, in turn, led to further biological changes. The development of language would have allowed *Homo sapiens* not only to couple suggestion with ritual activity, but also to describe the resulting anomalous experiences. Such verbalizations would have created oral traditions regarding the supernatural.

Even before the development of language, hominids probably had anomalous perceptions. Anecdotal reports, folkloric accounts, and folk traditions from many societies suggest that animals perceive anomalous events. A huge number of accounts from many societies describe dogs, cats, and horses acting as if they were aware of apparitions. Stories in many societies attribute precognitive and telepathic capacities to animals. One common narrative motif is that of a pet demonstrating distress at the time of its absent owner's death; this behavior is observed by humans before news of the death arrives. The literature supporting these claims is anecdotal but extensive.[67] We should not dismiss this type of data merely because it seems scientifically anomalous—after all, the presently unknown cognitive processes that cause humans to perceive spontaneous anomalous experience probably occur in other mammals as well. Strong evidence exists indicating that chimpanzees, baboons, monkeys, cats, dogs, and other animals hallucinate, implying that altered states of consciousness and hallucinations are a function of mammalian nervous systems.[68]

Because anomalous animal behavior is similar to that of humans, it is logical to hypothesize that the basic forms of folk religious belief are derived from pre-linguistic and early-linguistic forms of hominid anomalous experience. Australopithecines, *Homo erectus,* Neanderthals, and early *Homo sapiens sapiens* probably experienced apparitions, extrasensory perceptions, out-of-body episodes, and similar anomalies. The development of a self-concept (an in-the-body experience) would allow the incidence of out-of-body experiences that support belief in the soul. Seeing a deceased person naturally leads one to believe in life after death. Extrasensory and precognitive perceptions cause belief that there is a cosmic structure to the universe, that something exists that is beyond normal human perception. Some anomalous episodes also support belief in precognition, the ability to magically foresee the future. Religious beliefs based on anomalous experience could emerge even among early humans who had limited linguistic ability. This would explain the incidence of religious behavior among Neanderthals, whose linguistic ability may have been limited.

Various artifacts suggest that Upper Paleolithic humans linked ritual behavior with medical procedures. Paleolithic people, at various locations, bored holes in living individuals' skulls, leaving several hundred specimens as evidence of this nonfatal operation. They also cut out and perforated pieces of bored skulls after the individuals' death. They may have used these pieces as necklace amulets.[69]

Ancient cave art may establish a link between ritual and altered states of consciousness. Upper Paleolithic engravings, paintings, and sculptures are often in the deepest and least accessible parts of the caves. If these works of art were created for decorative purposes, they would have been close to entrances where they could be seen in the half-light. The same rock walls were repeatedly used, and images were covered over again and again. Such activities could not have been merely to produce ornamentation for an inhabited part of the cave. Apparently the art was created for ritual purposes. At burials Paleolithic people left many symbolic objects and substances, such as "bull roarers" and red ocher, which probably had magical or religious purposes. Flutes made of bird or cave bear bones have been found at different sites, supporting the argument that ceremonies were accompanied by music.[70]

David Lewis-Williams and Thomas Dowson argue that cave images are entoptic, that they have special forms related to altered states of consciousness. Entoptic phenomena are visual sensations derived from the structure of the optic system anywhere from the eyeball to the cortex. The universality of entoptic phenomena cause similarities in perceptions associated with certain altered states of consciousness. Lewis-Williams and Dowson suggest that South African rock paintings and Lascaux cave art were created or visited by people who were in an altered state of consciousness or were inspired by people experienced with such states.[71] They catalog the entoptic features found in the art. Although some of Lewis-Williams and Dowson's arguments seem speculative (they suggest that artists painted entoptic images while seeing them, for example),[72] the connections between caves, sensory deprivation, and altered states of consciousness is too strong to ignore.[73] People living in deep caves may have experienced the effects of a reduced sensory environment, including hallucinations. Sensory deprivation increases hypnotizability;[74] the creators and viewers of cave art may have been creating a condition in which they would experience mental images coupled with hypnoidal states, and this would enable the belief that such images were meaningful. Although Paleolithic skeptics who lacked hypnotic capacities might have considered ritual actions superstitious, rituals may have had therapeutic benefits that presented a survival advantage to those who were most suggestible.

Modern researchers have studied religious experiences under semicontrolled conditions. They use such methods as sensory deprivation and isolation in wilderness areas.[75] Paleolithic people would have been exposed to these conditions and therefore would have had equivalent experiences (labeled by modern respondents as "religious" or "mystical"). Anthropologists note that hunters and gatherers have more free time than do postindustrial people. Paleolithic hunters and gatherers with natural propensities could have trained themselves to experience anomalous perceptions just as modern hunters and gatherers do. It would be possible to conduct experiments to test hypotheses derived from the ritual healing scenario. People in isolated cave environments for extended periods of time are predicted to become more hypnotizable, to hallucinate, and to report greater frequency of other anomalous experiences than control groups.

D'Aquili proposes an evolutionary scenario that differs in some respects from the ritual healing theory. Although his theory is more physiologically precise, it includes features (such as "group selection") that have fallen out of favor among modern anthropologists. D'Aquili argues that rituals synchronize "affective, perceptual-cognitive, and motor processes within the central nervous system of individual participants."[76] He believes that ceremonial rituals promote intragroup cohesion by decreasing or eliminating intragroup aggression:

The core central experience of human religious ritual, when it works for an individual, is a marked attenuation of intragroup aggression and the experience of union or oneness. As with meditation, the experience of oneness is not further specified in the ritual effect itself. The most specific the experience is in itself is a vague sense of union with other ritual participants. Beyond this, what the oneness signifies and unites is expressed by the mythic system of meaning in which the religious ritual is embedded.[77]

D'Aquili suggests that the feeling of oneness and unity occurs when the ritual triggers activity in certain parts of the nondominant parietal lobe. He argues that repetitive auditory and visual stimuli such as drumming, chanting, and dancing drive cortical rhythms that can produce intensely pleasurable, ineffable experiences in humans. Ritual stimuli bring about "simultaneous intense discharges from both the sympathetic and parasympathetic human nervous systems."[78] During such meditative and ritual states, logical paradoxes and polar opposites appear both as contradictions and as unified wholes.

This experience is coupled with the intensely affective, "oceanic" experience which has been described during various meditation states as well as at certain nodal points of ritual. During intense meditative or ritual experiences . . . the experience of the union of opposites . . . is expanded to the experience of the total union of self and other, or, as it is expressed in the Christian tradition, the union of the self with God.[79]

In d'Aquili's view, humans are motivated to engage in rituals because of the pleasure they obtain from feeling special sensations. As a byproduct, they achieve a unity with their group that supports the development of religious ideologies. D'Aquili implies that groups that practiced such rituals prevailed over those that did not. As a result, religions arose from the "oceanic" experience derived from group rituals.

Although selection at the group level may have occurred, most evolutionists believe that selection is more common at the gene level.[80] Genotypes contributing to ritual behavior that allows groups to achieve unity would not necessarily have been passed on in greater number to future generations if individuals lacking the gene did not have reduced fertility. Also, humans over the centuries have found that the enlightenment experience, the powerful feeling of unity, is relatively rare. Because only a minority of individuals experience such intense mystical episodes, they probably would not be a source of unity for all members of a group. Finally, modern individuals who are thought to be enlightened tend to demonstrate a detachment that is associated with decreased fertility. This observation is consistent with studies of people suffering from temporal lobe epilepsy that is linked to increased religiosity—they are less interested in sexual activity.[81]

C. Parker reviews the weaknesses of group evolution models.[82] Group selection would influence evolution only if the gene contributing to group survival caused the group survival to be high enough to counteract the effects of deaths of people with the gene. If people can leave one group and join another, group selection models are not appropriate. But studies of the group dynamics of primates reveal that migration of individuals between groups is a regular feature. Although factors contributing to group benefit cannot be ruled out, models based on survival at the level of the gene appear more powerful than those based on survival at the level of the group.[83] Hypnotic rituals may contribute to group unity, but this function probably does not explain the original selection for hypnotizability genotypes.

Yet many of d'Aquili's arguments are consistent with those of the ritual healing theory. D'Aquili writes:

> Numerous reports from many religious traditions point to the fact that such [mystical] states yield a feeling not only of union with a greater force or power but also an intense awareness that death is not to be feared, accompanied by a sense of harmony of the individual with the universe. . . . When ritual works (and it by no means works all the time), it powerfully relieves our human existential anxiety, and at its most powerful it relieves us of the fear of death and places us in harmony with the universe.[84]

Activities that reduce anxiety provide survival advantages on the individual, or gene, level. On this issue, d'Aquili's position coincides with that of the ritual healing theory. But the ritual healing theory does not assume that Paleolithic people were swayed by their mystical experiences to create religion. Indeed, we shall see that the evidence regarding Paleolithic religion does not indicate that they practiced religion as we generally think of it.

Bruce Dickson points out that we should be able to gain insights into Paleolithic religion by analyzing material remains.[85] After all, archaeologists far in the future may gain insights into the religious systems of Europe during the last two millennia by analyzing artifacts. They may recognize religious buildings, mortuary remains, and artwork; identify and plot distribution of topologies and subtypes over time and space; and make inferences from these observations. They might conclude that Christianity had a hierarchical pantheon centered about one male; that it included a small number of subordinate, nurturing females; that its worldview focused on danger and violence (involving a crucifixion and other martyrdoms); that it included a well-developed belief in rewards and punishments following death. They would observe that virtually all artwork during medieval and earlier eras pertained to religion, and therefore they might regard the people as religious during those times. This hypothetical analysis gives us insights into the potential accuracy of the conclusions we might derive from our study of Paleolithic relics.

Paleoanthropologists note that the thousands of images of animals are naturalistic portrayals while the hundreds of images of humans are often merely sticklike silhouettes. Animal and human images tend to have no explicit relationship with each other.[86] Geometric depictions (dots, circles, cupules, dashes, rods, bars, and rectangles) arranged in various alignments suggest the perception of periodicity in the passage of time. Other images reflect sexuality: vulvas are common but phalluses are rare. On the basis of his analysis of this evidence, Dickson concludes that religion in Franco-Cantabria during the Upper Paleolithic period was:

> a) based upon a complex intellectual and theological order and b) was ultimately experiential in inspiration. Second, two profound natural phenomena—a) a perceived cyclicality in the passage of time and b) the dialectic of human sexuality, especially the periodicity and fecundity of women—were generalized into universal principles or "grand analogies" that formed the basis of speculation and thought about nature, humankind, the universe, and reality. This model of social and material reality was embodied and reflected in the great parietal art caves of Franco-Cantabria.[87]

Anthropology texts, such as the one by Marvin Harris, echo this analysis in broader terms: "The history and ethnography of art are inseparable from the history and ethnography of religion. As we have seen, the earliest paintings found on the walls of caves are generally assumed to have been painted as part of ancient religious rituals."[88] Harris goes on to note that puberty rituals, chanting, production of songs, body painting, dancing, and the making of masks, statuary, carvings, and other articles are intrinsically linked to religion in nontechnological societies. This evidence supports the proposition that art and religion were linked during the Paleolithic period.

But the Paleolithic evidence is unclear. Our interpretation of Paleolithic art may be clouded by assumptions that we derive from our observations of modern hunter and gatherers. Denis Vialou refuses to make inferences that are common within anthropology:

> Two bison-men engraved in the Sanctuary of Les Trois-Frères, Ariège, together with a third engraved in the older Magdalenian cave of Le Gabillou and a few rare beings with combined human and animal features, cannot provide sufficient evidence to substantiate the idea that shamanism underlies Paleolithic art. Similarly, the model of a bear riddled with blows from a spear that was found in the immense Magdalenian cave of Montespan, and a few other "wounded" depictions here and there, cannot justify interpreting Paleolithic art as having a religious or magical purpose.
>
> It is quite probable that magic rituals took place in these hunting communities . . . and also that shamans may have exerted religious powers on people and animals. Indeed, several clues found in burials and in settlements might lead to this conclusion, but this is simply not the case with their art.[89]

Vialou's argument seems sound. The ratio of images that suggest the influence of religion to images without any religious implication is very small. The ritual healing theory provides a perspective for interpreting this evidence. Paleolithic art and religion were probably linked, but as Vialou notes, the art does not compel that conclusion. Compared to medieval European artifacts, cave art images allow far less certainty that Paleolithic people were religious. If they were religious in the medieval European sense, their artwork would have better reflected that sentiment. Although they probably had shamans who conducted trance performances (as portrayed on cave walls in a few places) and devised ideologies as Dickson suggests, the average Paleolithic person may have been less capable of religious sentiment than were later humans. The ritual healing theory posits that Paleolithic rituals brought about an increase in the frequency of genotypes related to religiosity, allowing later societies to create artwork that was undeniably religious.

Historical Analysis of Ancient Medical Practices

One way to evaluate the hypothesis that ancient medical practices were linked to hypnotizability and religion is to analyze the earliest writings pertaining to medical treatment. If religion evolved from Paleolithic medical practice, we would expect all ancient societies to have devised medical systems based on religious ritual. Historical evidence supports this hypothesis.

The world's oldest existing medical manuscript is a small clay Mesopotamian tablet inscribed around 2100 B.C. Additional Mesopotamian texts were created during every century afterward. The largest Mesopotamian medical collection is the *Treatise of Medical Diagnoses and Prognoses,* about forty tablets that predict the future of the patient's condition on the basis of bodily, natural, and other signs. According to Guido Majno, "The actual prognostic significance of the work, in the modern sense, is about nil . . . the treatise is essentially a handbook of the sorcerer, not a manual of medicine."[90] Although the Mesopotamian pharmacy included about 250 medicinal plants, 120 mineral substances, and 180 animal and other drugs (mostly unidentified), the system was basically religious.[91] Its purpose was to reconcile the patient with the transcendental world. Henry Sigerist notes that, "in all civilizations of this area, religion, magic, science, and learning were one, an inseparable whole"[92] and concludes that "Mesopotamian medicine was psychosomatic in all its aspects."[93]

The author of the Smith papyrus, the first Egyptian medical text, lived sometime between 2600 and 2200 B.C. A copy of his text created in about 1650 B.C. has survived. The oldest existing Egyptian medical text, the Kahun papyrus, appeared around 1900 B.C. and was followed by half a dozen others. Like Mesopotamian medicine, Egyptian medicine had a religious basis, although it developed a secular component at an early stage, especially with regard to surgery.

Since spirit intrusion or at least affliction by a spirit—demon or ghost of the dead—was considered the chief cause of disease in the magico-religious medicine of Egypt, the means for prevention and cure consisted in the wearing of amulets, the reciting of incantations, and the performing of certain rites, all intended to keep off or drive out the spirits.[94]

Egyptian incantations have qualities that would ensure their effectiveness as psychosomatic medicine. They often included a kind of recommendation that consisted of marginal notes by a previous scribe who had used the charm and found it effective. These comments were incorporated into the text, increasing the expectation that the incantation would bring about the desired effect.

The recommendation may be very simple, consisting of a word or of a few words added as marginal notes by somebody who used the charm and found it satisfactory. . . . Later, as happened usually in such cases, the note became part of the text. Thus we find recommendations like "good," or "very good," or "really excellent, proved many times," or "I saw its good effect," or "really proved, I saw its effect on myself," and similar remarks undoubtedly made by the user of the charm. Simple as these words were they increased the suggestive power of the text.[95]

Ancient Greek medicine reveals a similar basis in hypnosis, religion, and the placebo effect. The only type of homeostasis beyond the compression of a wound and the application of poultices mentioned by Homer is the Ἐπαοιδή, which literally means "a song sung to" or "over [the wound]."[96] The Hippocratic collection was a derivation of existing religious treatment methods. Its text was written by different authors in the fifth or fourth centuries B.C. Hippocratic Greek physicians regularly offered sacrifices to the gods and took part in processions in honor of the god Asclepius, whom they considered the father of their healing art. The Asclepius mentioned by Homer was a mortal man, but he was deified within folk tradition over time. Followers of Asclepius and Hippocratic physicians placed importance on dreams and psychological factors for diagnosis and treatment. People with incurable or chronic diseases were encouraged to try methods based on the use of amulets or the interpretation of dreams.[97]

Asclepian temples were not the first in Greece to use dream incubation as therapy. Amphiaraos, the grandson of the wise diviner Melampus (about 1400 B.C.), was, like Asclepius, raised to the status of a god. At his temple at Oropos (Attica), supplicants made sacrifices, slept on ram skins, and sought enlightenment through dreams. A votive relief dedicated between 380 and 370 B.C. shows the god performing an operation on a youth's shoulder (apparently during a dream). A background image depicts a snake licking the reclining patient (perhaps the patient's dream), and in another background image, the patient has fully recovered.[98]

Ancient historians, such as Dicaearchus (d. 285 B.C.), Plutarch (A.D.

Healing god Amphiaraos (votive relief dedicated between 380 and 370 B.C.). The god performs an operation on a youth's shoulder, apparently during a dream. A background image shows the reclining patient being licked by a snake, and in another background image, the patient has fully recovered. (photo by the National Archaeological Museum, Athens, Greece)

47–120), and Pausanias (c. 170), describe an alternate method for inducing occult experiences. Supplicants descended into the ancient seer-physician Trophonius's cave in order to obtain a vision. They reported out-of-body and visionary episodes during which they learned the future. The experience was "a kind of shock treatment associated with sensory deprivation."[99] This method seems parallel to that probably discovered by Paleolithic cave dwellers who may have found that sensory deprivation leads to altered states of consciousness and visionary experience.[100]

The major religious branch of Greek medicine, based on belief in Asclepius, preceded the Hippocratic tradition but reached its zenith in the era during which the Hippocratic text was being transcribed. Patients slept in special temple rooms and hoped for a dream in which the god Asclepius (or his symbolic forms) would either heal them spontaneously or provide them information regarding treatment. Eventually, more than four hundred temples were dedicated to Asclepius, and some of them were still functioning in the sixth century A.D. Pausanias visited the Asclepian temple of Epidaurus in the second century A.D. and saw six tablets on which were engraved "the names of men and women who were healed by Asclepius, together with the disease from which each suffered and how he was cured."[101] Three of these tablets and a fragment of a fourth, which together contain seventy case histories, have been found. Written during the fourth century B.C. they are an important source regarding the types of people seeking healing at the temple and the nature of the cures that were performed.

The cures that Asclepius provided were often reflections of the medical practices of the era, only more miraculous.[102] The possibilities for hypnotic and placebo effects are clear: people expected the dreamed remedies to work, and their belief granted the remedies greater potential to be effective. Sigerist summarizes the types of persons and diseases described by the steles:

> They are, to put it briefly, people suffering from certain chronic diseases—for instance old sores, or chronic ulcers such as the varicose ulcers—which resist treatment but may sometimes heal rather suddenly without apparent reason. They are, furthermore, the large group of patients showing symptoms due to hysteria.[103]

The interpretation that Asclepian healings were a product of hypnosis has been common since the late nineteenth century.[104] Supplicants of Asclepius were exposed to inductive ritual environments. They trusted the

Relief of Asclepius healing reclining figure.
(photo by the National Archaeological Museum, Athens, Greece)

method, focused their attention on a specific procedure, sought information within an altered state of consciousness, perceived that their psychosomatic symptoms had been removed, and found that nonpayment of the offering could result in a return of the symptoms. These features coincide with the phenomena observed in other hypnotic therapies such as Mesmerism and modern hypnotherapy.[105]

> If a patient were hypnotizable, this unique combination of factors would surely increase susceptibility and a good trance state would seem very likely. And if the patient were not hypnotizable, these factors might combine to produce a subjective perception that there had been some improvement. From the standpoint of contemporary psychotherapy, the Asklipians combined cognitive, affective, behavioral, social, physical and spiritual treatment factors, a very powerful holistic combination.[106]

The theory that Asclepian healings involved hypnotic processes coincides with sleep research findings. Studies indicate that some subjects will comply with pre-sleep suggestions to dream about a specific topic.[107] In one controlled study, Kathryn Belicki and Patricia Bowers demonstrated that hypnotic ability was significantly correlated with the degree that dreams were changed by pre-sleep instructions.[108] H. B. Gibson found that women who enjoyed dreaming, who arrived at creative ideas while dreaming, and who reported that their future had been foretold in their dreams were more highly hypnotizable than women who failed to report this pattern.[109] People who are hypnotizable tend to report more subjective paranormal experiences such as precognitive dreams.[110] These findings resonate with the nature of hypnosis,[111] a process that involves the passage of information to and from the subconscious. They suggest that Asclepian cures were a form of lay hypnotherapy.

The earliest Indian and Chinese medical texts, which are more recent than those of Egypt and Mesopotamia, reveal equivalent patterns. Many folk healing practices were connected with religious doctrines. The spread of Buddhism from India to China was associated with the spread of magical practices and folk healing even though the Buddha himself placed little emphasis on miracles.[112] Historical data suggest that the most ancient medical practices in all regions were derived from rituals based on hypnotic, religious, and placebo-oriented procedures.

The Time Frame of the Ritual Healing Theory

The human evolutionary scenario allows sufficient time for ritual practices to have brought about profound physiological transformations. Ritual skull mutilations may have occurred by 150,000 B.P. and intentional burials as early as 68,000–78,000 B.P. Complex symbols, probably associated with linguistic ability, were used during the Upper Paleolithic period (beginning between 40,000 and 35,000 B.P.). Evolutionary history indicates that extremely rudimentary symbolization occurred for perhaps 150,000 years, rudimentary symbolization for about 70,000 years, and complex symbolization for over 30,000 years, sufficient time for these practices to have an impact on the development of hypnotic abilities that affected religious ideologies.

Charles Lumsden and Edward Wilson argue that the coupling of genes and culture drove the growth of human intellect forward at a rate perhaps unprecedented in the history of life.[113] They contend that a "thousand year rule" applies: the co-evolution of genes and culture could cause significant genetic changes within a mere fifty generations, or approximately one thousand years.

Although exponential models of change in genotype frequency are merely approximations, calculations reveal the plausibility of the proposed scenario. If shamanic treatments caused the incidence of genotypes related to high hypnotizability to increase by 2 percent each generation (a conservative estimate in light of modern hypnosis studies), genotype prevalence related to high hypnotizability would expand from 1 percent to the 15 percent found in modern populations in 138 generations (an estimated 2,760 years).

Several factors make modeling complex. Hypnotizability cannot increase limitlessly because its extreme forms are associated with psychopathology.[114] Furthermore, pre-literate societies use shamanic and hypnotic rituals for both benevolent and malevolent purposes; malevolent use could reduce the prevalence of hypnotizability genotypes.[115] Finally, a society's attitudes and values regarding hypnotizability affect the fertility of those who possess the trait (see the appendix for mathematical models regarding costs and benefits of hypnotizability). Even when we take these factors into account, however, the ritual healing scenario remains plausible because *Homo erectus* probably experienced group altered states of consciousness around fires 700,000 years ago and *Homo sapiens* probably sought healing by using shamanic and hypnotic rituals involving verbal suggestions for over 30,000 years. There was sufficient time for a modest genotype selection mechanism to have meaningful impact.

All of this evidence suggests an evolutionary progression from animal hypnosis to human therapeutic ritual and eventually to religiosity. First, animals reveal the "Totstell reflect" (animal hypnosis), a phenomenon that appears to be shaped by evolutionary processes. Chimpanzees demonstrate a hypnotic capacity similar to that of humans, as well as rudimentary forms of ritual and associated altered mental states. They also reveal slight capacity to provide support for the injured. Second, Neanderthals provided support for their injured and handicapped fellows, which demonstrates a cultural advance. Even though their language capacities may have been restricted, they revealed rudimentary capacities for rituals such as skull mutilation and intentional burial. Third, *Homo sapiens sapiens* gained the ability to use symbols and eventually language. Life in caves and wilderness environments is conducive to the development of altered states of consciousness, and humans devised complex rituals associated with ASC. Analysis of prehistoric art compared to European Medieval art suggests increasing religiosity over time. And finally, the oldest written medical texts from many societies link medical treatment with religious ritual, which indicates a hypnotic or placebo-based treatment of disease.

Unlike the prevalent theories explaining the origin of religion, these arguments are amenable to scientific evaluation. The existence of language, caring attitudes, and ritual altered states of consciousness among early humans would have selected for genes associated with hypnotizability. The existence of ancient texts connecting religious rituals with medical practice also suggests that humans used ritual healing for a sufficient time for it to have affected the frequency of related genotypes.

2
Fertility, Childbirth, and Suggestion

The video *Healers of Ghana* portrays a ceremony among the Bono people.[1] The narrator explains the actions of a healer, Kofi Ifa, who after ritual preparation, goes into trance and greets the people gathered to watch his performance.

> Narrator's voice: Greetings are important in Bono society, and the deity now arriving through Kofi Ifa must welcome his guests—among them, Kofi Duncor, Ifa's mentor. Like most witch-catching healers, Ifa can become possessed when away from his shrine.
>
> [The video shows Ifa greeting his mentor, then dancing while others drum. Ifa then takes an infant from a mother in the audience.]
>
> Narrator's voice: To demonstrate its strength and vitality, Kofi Ifa displays an infant, born of a woman thought to be barren before he applied an herbal cure.
>
> [The video presents images of Ifa holding the baby while in trance, then playing with it while the audience laughs appreciatively.]

All over the world spiritual healers claim that fertility can be enhanced by magical rituals. If some forms of infertility have psychosomatic bases, such practices would increase fertility, raising the frequency of genotypes related to suggestibility.

Carol Laderman observed Malay healers reciting a *jampi,* or incantation, during a woman's labor. The ritual was "meant to enhance the safety of mother and child by bringing them into a more harmonious relationship with the Universe, and, at the same time, keep the mother's mind on holy matters instead of on her pain."[2] Laderman observed a particularly difficult labor during which a healer was summoned:

> He knelt on a mat beside the laboring woman, his manner soothing rather than commanding, and his voice clearly audible. . . .

During his recitation, the woman's moans ceased, her expression cleared, she appeared to gather further strength, and her baby was born within a half hour of the jampi's end.[3]

Laderman's description of the rhythmic incantation implies that it served as an hypnotic induction—effective for hypnotizable people.[4] Malay spiritual specialists stress that success requires a melodious voice so that the patient can go into trance. Laderman notes that the prolongation of labor due to fear is associated with higher than normal perinatal mortality rates. She discusses the implications of her observations:

> The medicine man's incantation, by reducing the laboring woman's anxiety, may thereby reduce her epinephrine production. With her fears under control, her uterine inertia may be counteracted and the normal functioning of productive contractions restored. . . . This placebo effect may not be limited to the application of material substances but may be evoked as well by the power of words. In addition to reducing epinephrine production, the medicine man's incantations may also change his patient's biochemistry by encouraging her body to produce higher levels of endogenous opioids than ordinarily occur in the course of normal labor.[5]

Mothers who are more hypnotizable respond more fully to healers' inductions, and as a result, they are predicted to have fewer birth complications. Folklore accounts from many localities throughout the ages include mention of rituals and charms that reduce the pain and anxiety associated with childbirth. Loudell Snow interviewed Michigan respondents who told her that putting a knife under the mattress of a woman in labor reduces the pain:

> Olouise had experienced it herself and knew that it worked; the midwife never made her go through very many of these pains before she placed the knife under the mattress. . . . And the staff of a Lansing hospital maternity ward never suspected that Arlene Bauer's grandmother had a hatchet in her knitting bag when she came to visit her granddaughter. . . . [She] merely slipped it under Arlene's mattress when no one else was around. And did it work, I asked? Yes, it did. "It cut those pains right in half."[6]

Such symbolic actions are waking suggestions that reduce the pain of hypnotizable people. Trance is not required (chapter 4 reviews the attributes of hypnosis).

Does the medical literature support these arguments? Can psychological factors affect fertility and birth outcomes? Would rituals among hominids have increased fertility and reduced childbirth mortality?

The transition from *Homo erectus* to *Homo sapiens* probably occurred rapidly in evolutionary terms. According to the commonly accepted concept of *punctuated equilibrium,* many species are well adapted to their environment and hence are in equilibrium with it. During eras of equilibrium there is little evolutionary change. But if this equilibrium is thrown out of balance (by a change in climate, for example), rapid evolutionary change can occur, creating a new species.[7] A predominant scenario portrays *Homo erectus* individuals gaining advantages from communicating and cooperating with each other. This activity could have led to an unstable situation. Their environment and their mating behavior began selecting for genotypes associated with the capacity to symbolize and eventually for linguistic ability. As a result, *Homo erectus* developed large, intelligent brains, babies' head size increased, and childbirth became problematic.

We can only guess about the rate of ancient *Homo sapiens* mortality during childbirth. Judith Walzer Leavitt estimates that one in thirty American women died during childbirth in the 1880s (she assumes that women had, on average, five children, and that one maternal death occurred for each 154 live births).[8] Using women's letters, Leavitt documents the high level of fear and anxiety that this level of childbirth mortality produced among American women in the 1800s. UNICEF data for sub-Saharan Africa indicates an average of 980 maternal deaths per 100,000 live births (one death per 102 births).[9] Sierra Leone, Somalia, and Guinea exceed this average, with more than one death for each 63 births, a rate more than twice that of the U.S. 1880s estimate.

Although hominid life was not similar to life in modern Africa, hominid childbirth may have been even more dangerous, particularly during times of famine and considering the increasingly large newborn head size associated with the evolutionary transition from *Homo erectus* to *Homo sapiens*. Childbirth must have been extremely stressful for these individuals, and psychological factors probably played a role in childbirth mortality. As cultures became more complex, women probably found themselves in less powerful positions and experienced associated social stress. About two adult male Paleolithic burials have been found for each female burial, suggesting early gender stratification.[10]

Much evidence indicates that stress affects rates of conception, spontaneous abortion, and birth outcome.[11] If rituals reduced the effects of stress, these practices would have had evolutionary impact. The relationship between ritual and fertility is of particular interest because it implies direct selection of genotypes associated with suggestion.

Genetic selection related to childbirth has caused differences between men and women: modern women tolerate higher levels of pain than men. Although men may have benefited from ritual treatments following hunting and warfare injuries, these events cannot be anticipated to the same degree as childbirth. If therapeutic rituals benefited females more than males, this might explain gender differences in religious sentiment, possible gender differences in rates of anomalous perception,[12] gender differences in temporal-lobe functioning, and the existence of female statues and other artifacts that suggest that Paleolithic rituals involving fertility and childbirth were particularly important.

Although this chapter focuses on fertility and childbirth, I hypothesize that therapeutic rituals had evolutionary impact on death rates due to many other problems as well. Relationships between stress and infectious disease,[13] coronary heart disease,[14] peptic ulcer disease,[15] and hypertension[16] have been well documented. Although these disorders were probably less prevalent among Paleolithic groups than among modern people, all diseases and injuries affecting humans during the childbearing years may affect fertility rates and distributions of genotypes. Any treatment system that was used for thousands of years would have had evolutionary impact if it was more effective for certain people. Although conclusions from individual studies are not always clear-cut, evidence indicates that psychological factors affect fertility and health in general and that treatments enhancing psychological heath would have had evolutionary impact on rates of infertility, spontaneous abortion, and childbirth complication.

Infertility

There is much evidence that psychological factors affect fertility in animals. Studies of both rats and rhesus monkeys indicate that stress elevates cortisol levels, affecting the production of ovarian steroid hormones.[17] Rhesus monkeys[18] and Cynomolgus monkeys[19] whose social status was lowered experimentally demonstrated suppressed levels of reproductive hormones.

This relationship also exists in humans: "There is an overwhelming consensus among mental health professionals and experts on reproductive dysfunction that psychopathology can psychosomatically cause or contribute to subfecundity [reduction of fertility] in both men and women."[20] If there were no relationship between stress and fertility, that lack would constitute an anomaly because psychosomatic disorders appear in all other systems of the body. Reproductive organs are thought to be common targets of psychosomatic processes. The debate regarding psychosomatic effects on humans' reproductive organs concerns not whether such effects occur but their impact on fertility rates. Infertility specialists need to know how intensely they should search for psychological problems before they can be satisfied that such problems are not a major component in a particular patient's infertility.[21] Infertility rates are positively correlated with stress, but the evidence is far more complex than might be supposed.[22]

Joseph McFalls has attempted to determine the degree to which psychological factors affect infertility. He estimates that 18 to 30 percent of infertility has a psychic-stress component. He believes that stress has a greater impact on male fertility than on female fertility; it causes perhaps 2 percent of males to suffer from impotence, while only 1 percent of females are unable to engage in intercourse due to vaginismus and dyspareunia (pain during intercourse).[23] Groups undergoing high stress would experience higher incidence of infertility.

More recent studies allow better understanding of the parameters affecting

McFalls' estimates. The research literature pertaining to conception covers four areas (which are not mutually exclusive): the labeling of psychosomatic infertility, comparison of patients with and without somatic infertile conditions, psychoneurophysiological or endocrine mechanisms, clinical studies involving care of infertile patients.[24]

The Labeling of Psychosomatic Infertility   Some patients' infertility remains unexplained after the completion of all available diagnostic tests.[25] Longitudinal studies of women whose infertility was unexplained by diagnostic tests reveal that many later conceived. Overall pregnancy rates as high as 50 percent have been reported.[26]

Although this observation suggests the validity of the category "psychosomatic infertility," the practice of labeling infertility psychogenic after excluding organic causes is clearly inadequate.[27] Advances in scientific understanding allow greater assignment of apparently psychogenic cases to organic categories, rendering the number of cases in the category "psychosomatic infertility" smaller. On the other hand, scientific discoveries are revealing that psychological factors can affect physiological functions, potentially broadening the category.

The assertion that psychological factors can be a primary cause of infertility has not gone unchallenged.[27] Various studies of patients labeled as psychosomatically infertile are methodologically weak, and some common assumptions are invalid. Infertile women who adopt infants are not more likely to conceive afterward as has been claimed by some researchers. Although other evidence supports the assertion that stress can cause infertility, the line of research based on women whose infertility remains unexplained has not completely established the connection between stress and infertility because the application of the label of psychosomatic infertility is problematic.[28]

Comparison of Patients with and without Somatic Infertile Conditions
Various studies compare psychological data gathered from samples of female patients with organic infertility to samples of women whose infertility lacks an organic explanation. Psychological test scores of functionally sterile women differ from those of mechanically sterile women.[29] Differences in personal histories indicate that functional sterility can be a protective mechanism.[30] J. C. M. Wilkinson and associates observed that "women with pelvic pain with no obvious organic cause have certain psychological characteristics which clearly distinguish them from normal women without pelvic pain. They also differ to a lesser extent from women with pelvic pain due to organic cause."[31] Women with functional sterility differed from mechanically sterile women in four dimensions of personality; their most pronounced difficulties were in the dimensions of the mother-daughter relationship and in their acceptance of the feminine role.[32]

Yet many of these studies fail to take into account the stress that may result from unexplained infertility;[33] unexplained infertility may be the cause, rather than the product of, emotional problems. A. A. Templeton and G. L. Penney reviewed case records of five hundred consecutive couples whose infertility remained unexplained and found no differences between them and patients with a detectable, organic cause for their infertility.[34] Studies that lack a control group cannot conclusively determine whether there is a causal relationship between stress and infertility.

Psychoneurophysiological or Endocrine Mechanisms   Various physiological explanations support the assumption of a causal relationship between stress and subfecundity. The mechanisms acting on the hypothalamus from superhypothalamic centers and the correlations between the various neurotransmitters appear to mediate the relationship between fecundity and psychological processes. When people experience emotional tension, they produce serotonin, which inhibits the antiprolactin system. The increase in prolactin can affect fertility in women and sexual function in men.[35]

Psychological factors have direct effects on ovulation in other ways as well. A hypothalamic-pituitary block generated by fear or stress may inhibit production of gonadotropins by depressing the secretion of the LH-releasing factor in the hypothalamus, and therefore inhibiting ovulation.[36] Like prolactin and serotonin, catecholamines, adrenal steroids, and endorphins, all affect ovulation and in turn are affected by stress.[37] Emotional factors act also on immunological processes, and particularly on antigens and antibodies in the cervix and in sperm.

The autonomic nervous system can also affect fecundity by controlling the responses of the involuntary muscles that are found in most of the reproductive organs (the fallopian tubes, uterus, vagina, sperm ducts, penis, and so on). For example, the muscular spasms that occur in the sperm ducts of men who are under stress can interfere with semen transmission. Psychological factors that affect the autonomic nervous system can also affect the responses of the organs of secretion (the seminal vesicles, prostate, cervix, and so on). For example, women may secrete abnormal cervical mucus that is hostile to spermatozoa, thus reducing conceptive ability.[38] Experiments suggest that the number of spermatozoa reaching the ampulla of the fallopian tube is decreased in ewes that show signs of discomfort or fright.[39] In males it is thought that a stress-induced increase in epinephrine causes a decrease in oxytocin production, which then affects sperm transport in animals.

Clinical Studies involving Care of Infertile Patients   Clinical studies of infertile patients have identified relationships between psychosocial stress and fertility. Two studies link psychological stress with poor outcome after *in vitro fertilization* (IVF).[40] Additional studies indicate that women who are more anxious and women expressing negative emotions have lower pregnancy rates after IVF.[41]

Vaginismus, the syndrome in which a woman unconsciously and spastically contracts components of the pelvic musculature, is treatable through psychotherapy and counseling. With full cooperation from both members of the sexually dysfunctional marital unit, William Masters and Virginia Johnson have achieved 100 percent treatment success.[42] Richard Kornhauser and his associates report success in treating impotence with brief psychotherapy.[43]

Major studies indicate that hypnosis is useful for treating psychosomatic sexual disorders in men. Harold Crasilneck reports on the use of hypnotherapy in the treatment of 1,875 males for psychogenic impotency over twenty-nine years. He concludes that "hypnosis should be considered as a primary treatment modality when those cases of psychogenic impotency do not respond to the pharmacological regimen."[44] There are many case reports of the use of hypnotherapy for the treatment of psychosomatic sexual dysfunction, but more controlled studies are needed.[45]

Although critics point out gaps in the literature, existing studies do indicate that stress reduces fertility and that therapy can reduce the effects of stress. This evidence supports the argument that spiritual healing (indigenous psychotherapy) can affect fertility. Spiritual therapies within small indigenous groups are often more effective than Western psychotherapy.[46]

Spontaneous Abortion

The possible linkage between stress and spontaneous abortion, or miscarriage, points to another way in which ritual healing may affect fertility. John Rock estimates that the rate of spontaneous abortion lies between 30 percent and 50 percent of all pregnancies. Although about half of these miscarriages can be attributed to chromosomal abnormalities, a relatively high percentage may have a psychological origin.[47]

Edward Mann observed that patients who report repeated miscarriage:

> have been found to be remarkably alike in many emotional respects. Perhaps the most striking similarity is to be found in their tendency to react somatically to psychic situations. . . . [They] usually present lifelong histories of psychosomatic reactivity which they themselves associate with certain types of emotionally stressful situations.[48]

Controlled studies have established a relationship between psychological factors and habitual miscarriage. W. H. James examined outcomes of pregnancies in a randomly chosen group of women who had previously experienced three spontaneous abortions. All these women received psychotherapy. He calculated that without psychotherapy less than 42 percent of the pregnancies whose outcomes he examined would have resulted in live births. In contrast, 80 percent of pregnancies resulted in live births after psychotherapy. As a result, James endorses the beneficial effect of psychotherapy for reducing repeated miscarriage.[49] Carl Tupper and Robert Weil investigated thirty-eight women who had previously experienced three consecutive early pregnancy losses. The viable pregnancy rate in the patients who underwent psychotherapy was 84 percent compared to 26 percent in the control group ($p < 0.01$). Tupper and Weil recommend a psychotherapeutic approach in the treatment of patients who suffer repeated fetal loss.[50]

Childbirth Complications

Grantly Dick-Read pioneered the practice of using relaxation techniques in labor. He based his theory on the concept that fear results in muscle tension, which inhibits the normal dilation of the cervix, and that anxiety leads to disregulation of muscles, causing muscle spasms, hypersensitivity, and contraction of vessels. He described this condition as the fear-tension-pain syndrome. Dick-Read opposed the use of risky anesthetic procedures.[51]

Although recent researchers have refined and modified Dick-Read's theory, it has become axiomatic that psychosocial and emotional factors such as anxiety and stress are related to abnormalities of pregnancy, parturition, and infant status.[52] Many studies support this understanding.[53] N. L. Juznic, L. Vojvodic, and D. Avramovic write:

> Obstetrical experiences reveal that rather severe labor pain produces greater fear and agitation of the patient, provoking an increase in duration of labor, as well as a higher incidence of complications both in mothers and children. According to our experience, the psychological methods of training undoubtedly produce positive results. A more rapid opening of the uterine aperture, better oxygenation, and shorter delivery reduce the incidence of obstetrical complications.[54]

They cite a study by F. Stahler, E. Stahler, and R. Gutanian that notes a difference in postnatal mortality of 2.8 percent in a group of untrained patients compared to 0.8 percent in a trained group, and a difference in the postnatal morbidity of 20 percent compared to 6.4 percent.[55]

Similarly, Hans Molinski writes:

> Every obstetrician is familiar with patterns of delivery behavior, which are direct correlates of anxiety: either uncoordinated or loud behavior in general, for example screaming and kicking, or behavior that is characterized by aimless and inexpedient use of a quite often cramped musculature. The woman does everything wrong; she is unable to cooperate with the doctor and his assistants. She may bear down too early—before the baby's head is sufficiently advanced, to cite only one of many possible examples.[56]

Various studies indicate that training programs facilitate childbirth by reducing the incidence of pain, premature delivery, and childbirth complications.[57] Although controlled experiments have not proven any single approach to be superior, the most prevalent method presented in training, the Lamaze method, incorporates self-hypnosis techniques.

Modern theorists have devised sophisticated explanations for the means by which psychological factors affect childbirth. L. Zichella and associates review hypothesized mechanisms.

> It is well known that stress is associated with the increased secretion of catecholamines and 17-hydroxycorticosteroids at peripheral level, and in recent years various authors have examined the role of the catecholamines as possible mediators between stress and abnormal uterine functioning. It has been shown that women with abnormal labor have higher levels of catecholamines than their normal counterparts . . . and that emotional stimuli such as fear increase the motor activity of the uterus. . . . Studies on the effects of infusions of catecholamines on the uterus in vivo have shown that adrenaline in low doses inhibits spontaneous as well as oxytocin-induced human uterine contractions whereas noradrenaline promotes an increase in human uterine motility and at high rates, uncoordinated activity and hypertonus. . . . Thus there exist indications that catecholamines may act as mediators between stress and abnormal uterine action.[58]

Norman Morris also reviews research pertaining to mechanisms that connect psychological factors to complications of childbirth: "It is well known that during stressful episodes, both cortisol and prolactin are released in increased amounts in human beings. Various studies have been carried out in relation to the effect of adrenaline and noradrenaline upon the intact human uterus in late pregnancy and during labor."[59] Yet Morris does not find that a relationship has been established between high anxiety and the development of prolonged labor, incoordinate uterine action, fetal distress, the need for cesarean section during labor, the use of forceps delivery, or the condition of the fetus as demonstrated by scores on the Apgar tests that are administered shortly following birth. The failure of researchers to replicate previous studies indicates that the relationship between stress and birth complications is not as simple as might be assumed.

Marci Lobel's review found that studies employing multidimensional measures of prenatal stress indicated significant adverse effect on birth outcome.[60] Different birth outcomes appear to be associated with unique sets of prenatal psychological factors in the mother: if stress creates high levels of the maternal catecholamines that precipitate labor, then acute stress may result in preterm delivery. Chronic stress may result in impaired caloric utilization and low birth weight.

Pathik Wadhwa and associates cite a huge volume of studies that "consistently found that psychosocial factors are significantly related to the incidence of adverse birth outcomes."[61] These studies coincide with a large body of animal research

> which support and extends the conclusions of the above human studies. Experimental studies in rats, sheep, and primates suggest that prenatal stress is causally associated not only with adverse birth outcomes such as preterm birth and low birth weight, but also with adverse long-term neurodevelopmental outcomes related to brain morphology, physiology, and behavior.[62]

Their study suggests a future research direction:

> The present findings are consistent with the premise that maternal-placental-fetal neuroendocrine parameters are significantly associated, both in magnitude and specificity, with features of maternal psychosocial functioning in pregnancy despite the systemic alterations associated with the endocrinology of pregnancy. These findings provide a basis for further investigations of the role of the neuroendocrine system as a putative mediating pathway between prenatal psychosocial factors and birth outcome, and possibly also as a mechanism linking features of the maternal psychosocial environment to fetal/infant brain development.[63]

There is a body of research literature that indicates that psychological treatment systems, particularly hypnosis, can reduce both maternal anxiety and birth complications. T. M. Harmon and his associates found that subjects in a "hypnotic suggestion" group and highly hypnotizable women in a control group reported less childbirth pain.[64] Hypnotically prepared mothers' deliveries involved shorter stage-one labors, less medication, higher Apgar scores, and more frequent spontaneous delivery than control subjects' deliveries.[65] Controlled clinical studies also indicate that hypnosis can be effective for the treatment of preterm labor,[66] for stimulating fetal movements in anxious pregnant women,[67] and even for inverting the fetus from breech to vertex presentation.[68] These studies portray an almost magical quality: a life-threatening situation (a breech birth) can be alleviated by the use of hypnotic suggestion. This body of evidence supports the argument that spiritual healing techniques based on hypnotic processes can reduce the incidence of childbirth complications, thereby increasing survival rates of hominids.

Unfortunately, the present state of knowledge about physiology does not allow us to specify the exact mechanisms by which hypnotic suggestion affects unconscious processes. No studies show precisely how hypnotic suggestion affects maternal-placental-fetal neuroendocrine parameters, for example. Although hypnotic suggestions are thought to influence a wide variety of physiological functions that are outside of conscious control, most well-controlled experimental studies have focused on pain control.

Folklore accounts support the argument that ritual suggestions can reduce hemorrhage. Because on the surface the main cause of childbirth mortality is loss of blood, it might be helpful to compare folklore accounts of "bloodstoppers" who magically stop a wound from bleeding to controlled clinical experiments that reveal that hypnotic suggestion can reduce hemorrhage during surgery. Although blood flow is one of the many processes affected by hypnotic suggestion,[69] the stoppage of bleeding may have been particularly important during Paleolithic eras because hunting accidents, wounds from warfare, and postpartum hemorrhage were probably major causes of mortality.

North American "bloodstopper" accounts describe a magical practitioner uttering a spell that stops the patient's bleeding:

> The man or woman known as a "bloodstopper" recites a magical verse as a curing technique; he or she may be called upon for problems ranging from bleeding from minor cuts to postpartum hemorrhage. . . . The most commonly reported such verse is found in the Bible (Ezekiel 16:6): "And when I passed by thee, and saw thee polluted in thine own blood, I said unto thee when thou wast in thy blood, Live; yea, I said unto thee when thou wast in thy blood, Live."[70]

Bloodstopper stories come from all over the world. Wilasinee, the Thai healer I described in the introduction, punctured her skin with needles, but bleed only briefly. I observed similar bloodstopping feats in Sri Lanka (see chapter 3). The Greek poet Homer (ninth century B.C.) described treatment of Odysseus's knee wound, which was inflicted by a boar: "Autolykos' tall sons took up the wounded, working skillfully over the Prince Odysseus to bind his gash, and with a rune they stanched the dark flow of blood."[71] Their "rune" was a kind of song, a magical spell sung over the wound that would act as a hypnotic suggestion. Ancient people often bled to death from wounds or childbirth complications, but rituals were a method of treatment.

The folklorist Richard Dorson found it easy to collect bloodstopper accounts in Michigan: "Bill Johnson shoved a cork into a pop bottle, the gas burst the bottle, and the glass cut the artery on his hand. I went and got one of the girls and she passed her hand over the cut and the bleeding stopped. Then in the night it started again, and she came and did the same thing, and it never bled again."[72]

Clinical research supports the argument that ritual suggestion can reduce hemorrhage. Controlled studies indicate that blood loss during surgery can be reduced by the use of hypnotic suggestion.[73] E. Disbrow, H. Bennett, and J. Owings provided test group subjects with suggestions regarding relaxation and the importance of blood conservation (they suggested that "the blood will move away" from the area of surgery during the operation). The instruction group had significantly less blood loss during the spinal operations than did control or relaxation groups.[74] R. Hart used taped hypnotic induction procedures to aid the recovery of surgery patients. Patients in the experimental group required transfusions of significantly less blood than did patients in the control group.[75] D. Rapkin, M.

Straubing, and J. Holroyd exposed patients to hypnotic suggestion and found a negative correlation between hypnotic ability and blood loss during surgery; they demonstrated that the result was due to hypnotizability rather than to the placebo effect.[76] Hypnotic control of bleeding in hemophiliacs is well-known among hypnotists, particularly dentists.[77] Studies indicate that, in hemophiliacs, hypnosis can effectively reduce both the amount of factor concentrate required to control bleeding and the general distress level as measured by a symptom checklist.[78]

Folklore narratives are not exactly equivalent to controlled studies of surgical blood loss. The events described in folklore narratives exceed the limitations of normal hypnotic explanation. Some informants describe bloodstoppers who apparently affect bleeding from a distance, a phenomenon that cannot be attributed to suggestion. Bloodstoppers who are summoned to resolve an emergency may act like they are certain that the bleeding has stopped, demonstrating a kind of extrasensory perception. Some informants describe healers who stop the bleeding of injured animals, feats that also exceed explanation based on the hypnosis theory. An example combines these motifs:

> George MacDonald [the bloodstopper] had a neighbor whose horses cut themselves to pieces on a barbed-wire fence. The neighbor, Dan Trumbauer, ran to George, who said, "You go back, I'll be there in a few minutes as soon as I get dressed. They ain't a-bleedin' at all." When Dan got back, not a drop of blood was flowing from the horses' cuts.[79]

The disjunction between folklore claims and scientific assumptions exists in all types of healing narratives: some healers are thought able to cure someone at a distance without using suggestion. Even if alternate processes besides hypnosis explain these accounts (paranormal forces, spirits, memory and perceptual distortions, sleight-of-hand, and so on), wondrous-healing stories generate expectations that make the incidence of future hypnotic and placebo effects likely. Hypnotizable women may be more likely to deliver their babies safely as a result of hearing these narratives.

In all probability, the processes by which ancient human rituals selected for hypnotizability genotypes were complex. Shamanic treatments of infertility, spontaneous abortion, and childbirth complications were merely a few of the domains within which these genotypes were probably selected. The field of psychoneuroimmunology has uncovered some of the mechanisms that explain these processes, and it promises to shed more light with further research. With time, we can expect the development of better physiological models that describe how ritual healing and psychological factors affected human evolution.

3

The Anthropology of Wondrous Healing

When I was living in Okinawa, my friend and translator Takamiyagi helped me interview spiritual healers.[1] World War II disrupted Okinawan shamanic traditions, and as a result many new religions had emerged.[2] Each week Takamiyagi and I talked with another practitioner. We visited a Chinese New Age healer, a woman who had founded an innovative shamanic religion, a traditional Okinawan shaman, a medical doctor who believed he could photograph evil spirits, and a businessman who had founded a pseudoscientific healing cult.

Takamiyagi's experience as a hypnotherapist led him to believe that all spiritual healing methods were based on hypnotic processes. He assisted me partly because he hoped to pick up new "tricks" he could use in his own practice. One Saturday he called me on the telephone. "Meet me at the Pepsi-Cola bottling plant tomorrow morning!" he requested. "I have found another new religion. They do incredible psychic healings but they only meet on Sunday mornings."

On Sunday I learned that Takamiyagi had located an Okinawan charismatic Christian church. This was the first Christian service Takamiyagi had ever attended and he knew virtually nothing about Christianity. His summary of their beliefs, gained from listening to the preacher speak, amused me. "Apparently, this religion is not based on honoring one's ancestors—they don't believe as the shamans do that one's deceased relatives can cause problems," he informed me. "They believe that by placing their faith in one man, Jesus, they can drive out the demons that bring problems."

As the service continued, the preacher asked that the congregation join him in prayer. Takamiyagi became particularly attentive. "Look," he whispered. "They are going into trance and their leader is providing them with hypnotic suggestions. Soon we will see the results!" he prophesied.

"They are not in trance," I whispered back. "That's merely a cultural convention. It's the prescribed posture."

"That doesn't matter," Takamiyagi whispered. "They are following suggestions; they have adopted the trance position and even without trance, many will respond." Sure enough, members of the congregation were "filled with the Holy Spirit." They healed each other by laying on of hands, and many appeared to be in trance.

Later, the preacher described events he believed demonstrated that their faith was valid. They had cast out demons, witnessed miraculous healings, and prevailed in the face of difficult situations. He told of his original conversion to a staid form of Christianity in Kyoto, Japan, then how he had become a fervent believer in Okinawa as a result of his reading of the Bible and his personal experiences. He had overcome an emotional crisis not unlike the psychological sicknesses that shamans often describe.

One of the preacher's accounts was particularly amazing. A woman had come to him who had visions that always came true. Her most recent vision predicted that she would die. She consulted a shaman who told her that she must become a shaman herself in order to survive. When the preacher touched her, she shrieked with pain and fell to the floor, unconscious. The preacher believed that she was possessed by a demon. She began responding in a man's voice, saying: "You can't do it. You can't make me leave!" The preacher attempted for hours to exorcise the demon, demanding that it leave in the name of Jesus. He claimed that at one point the woman levitated above the bed. In the morning the demon departed, and the woman became a Christian.

Takamiyagi attributed these perceptions to hypnotic processes. He believed that both the woman and the preacher socially constructed these unusual accounts as a result of experiences derived from inferred suggestions. Takamiyagi voiced a notion common within the social sciences. He argued that social processes shape people's experiences and beliefs—and that hypnotic effects are a part of these social processes. "These Christians use hypnosis just like everyone else," Takamiyagi exclaimed. "They are similar to Okinawan shamanists in that they use trance behavior to create spirits. But they are different from shamans in many ways. They treat psychological problems as exterior to the family. I can see how this can be therapeutic for some people. I have a client who might benefit from this."

I tried to explain to Takamiyagi the theories that a cultural anthropologist might use to explain spirit possession.[3] Most anthropologists have little knowledge of hypnosis. They often focus on the group functions that rituals fulfill, examining the ways in which ritual practices reflect and support a particular culture. Takamiyagi found it hard to understand how this orientation explained much of anything. "Don't anthropologists understand what people actually feel?" Takamiyagi asked innocently.

Although anthropologists examine the meaning that local people place on occult experiences, they rarely seek those experiences themselves, I explained. This tendency, coupled with their focus on culture, contributes to unnecessarily abstract conclusions.[4] One anthropological reviewer noted that various studies demonstrate that possession trance is related to self-hood and identity, "challenging global political and economic domination, and articulating an aesthetic of human relationship to the world."[5] Such research involves macro-level analysis focusing on the relationship between spirit possession, colonialism, and power,[6] and results in "responses to changing social and historical circumstances."[7] Anthropologists have devised these formulations because they often see powerless people become possessed by powerful spirits in a manner that grants status.[8]

"I don't think these Christians are challenging the political or economic structure," Takamiyagi noted. "They appear to be above average in status and wealth. I think that the leader gains converts because his notion of spirits resonates with modern Okinawan culture. Just as shamans are possessed by ancestral spirits, these Christians are possessed by the Holy Spirit and sometimes by demons."

While Takamiyagi was impressed with the similarities between charismatic Christianity and shamanism, I was surprised at how similar the Okinawan exorcism was to accounts of Western witch hunters during the Renaissance. In both cases demons "proved" that they were authentic by performing paranormal feats, then were overcome by the power of Jesus.

To what degree are Takamiyagi's observations correct? Do all religions have a basis in hypnosis? Are the demon possessions reported in Okinawa equivalent to those described by Western Christians? Anthropologists attempting to interpret data gathered from a variety of cultures note that the concept of possession has different meanings in different societies. Christian Okinawan demons are not exactly equivalent to those in Renaissance Europe. The logical and philosophical meanings attributed to terms such as *sorcery, magic, possession, religion,* and even *science* are culturally specific. Some anthropologists argue that universal definitions for terms such as *possession* "cannot possibly work across time and space."[9] When anthropologists confine their discourse to the cultural arena, such assertions appear valid because all cultures differ.

Yet the ritual healing theory seeks to explain why so many similarities exist in all folk traditions even though interpretations vary. Universal elements suggest physiological bases. I argue that Western Christians of the Renaissance, present-day Okinawan Christians, and shamanists reveal experiential similarities because there is a physiological basis within hypnotic processes. We can investigate this hypothesis by comparing experiential accounts from a wide variety of cultures to determine whether recurring elements coincide with elements associated with hypnosis.

Some of the most interesting evidence bearing on this hypothesis comes from anthropologists' personal anomalous accounts. Anthropologists who have had anomalous experiences believe that these experiences grant them special insights regarding how indigenous peoples maintain their religious beliefs.[10] Edith Turner, for example, was participating in a Ndembu cere-mony (in Zambia) when she saw an apparitional substance thought to be responsible for the illness of a villager named Meru:

> Suddenly Meru raised her arm, stretched it in liberation, and I saw with my own eyes a giant thing emerging out of the flesh of her back. This thing was a large gray blob about six inches across, a deep gray opaque thing emerging as a sphere. . . . We were all just one in triumph. The gray thing was actually out there, visible, and you could see Singleton's hands working and scrabbling on the back—and then the thing was there no more. Singleton had it in his pouch, pressing it in with his other hand as well.[11]

Turner does not emphasize the social functions fulfilled by the cere-mony. Instead, she believes that she witnessed an anomalous event:

> In a book entitled *Experiencing Ritual* (1992) I describe exactly how this cura-tive ritual reached its climax, including how I myself was involved in it; how the traditional doctor bent down amid the singing and drumming to extract the harmful spirit; and how I saw with my own eyes a large gray blob of some-thing like plasma emerge from the sick woman's back. Then I knew the Africans were right, there *is* spirit stuff, there is spirit affliction, it isn't a matter of metaphor and symbol, or even psychology. And I began to see how anthro-pologists have perpetrated an endless series of put-downs as regards the many spirit events in which they participated—"participated" in a kindly pretense. They might have obtained valuable material, but they have been operating with the wrong paradigm, that of the positivists' denial.[12]

My argument is not that Turner is correct in arguing that "spirit stuff" exists[13] but that anthropologists who experience anomalous events often come to understand indigenous people's beliefs. The physiological basis for occult experiences, possession, trance, and related perceptions produces re-curring features within accounts of such events. These episodes, which in-clude ESP, sightings of apparitions, out-of-body experiences, psychokinesis, and sleep paralysis, appear to be related to hypnotic processes. Anthropo-logical observation can help us to determine whether they are indeed re-lated to hypnosis.

Unfortunately, most anthropologists are unaware of the nature of hypno-sis. Without firsthand experience, it is difficult for them to evaluate the claim that indigenous rituals use hypnotic processes. Yet knowledgeable observers see many connections between shamanism and hypnosis. They note that shamans and highly hypnotizable people have similar characteristics, follow similar methods of induction, and report similar experiential phenomena.[14]

Anthropologists' accounts illustrate this process. Bruce Grindal partici-pated in a Sisala death divination ceremony (Ghana). He saw a corpse be-came radiant with light, then dance and play the drums. But he apparently is unfamiliar with the hypnosis literature; he provides an explanation that is shrouded in anthropological vagueness:

> What I experienced, I now believe, was a synesthetic integration of my senses, whereby I perceived the rhythms of the music, the movement of the feet, the light of the campfire, and the hidden presence of the dead buried below combined in bright, yellow-white fibers of motion. I witnessed the collective power of the drummer's household and the whole lineage of the chief of Tumu. I experienced what the others around me experienced: the passionate resurrection of the power of the ancestors.[15]

Events such as the one Grindal describes cause indigenous people to believe in the power of magical rituals and in the reality of life after death. Lay people rarely attribute such episodes to "a synesthetic integration of the senses" although an agnostic hypnotist might concur with that opinion. Grindal's argument and similar contentions might be labeled "rationalist" because they provide a "rational" explanation for lay people's religious experiences.

Apparitional episodes like those that Grindal and Turner describe probably involve hypnotic processes. Grindal and Turner may be more hypnotizable than most people, and the setting and the ritual music and dancing would have facilitated their experiences. Hypnotizability allows people to have anomalous experiences that support beliefs which facilitate healing. Unlike most functionalist or rationalist explanations, this hypothesis is subject to empirical evaluation. I predict that spiritual healing outcomes and the frequency of unusual ritual experiences will be correlated with hypnotizability (measured by standardized tests) and cognitive openness (measured by questionnaires). Evidence already exists supporting these assertions. Recurring observations within the literature indicate that spiritual healing can be effective, that spiritual healing is most useful for treating ailments with psychosomatic components, and that spiritual healing is effective, in part, because of hypnotic and placebo processes. I will review evidence from the anthropological literature and from my own studies supporting these hypotheses, all of which can be evaluated by others.

The Efficacy of Spiritual Healing

Anthropologists have observed effective spiritual healing in a wide variety of settings for many decades.[16] The evidence suggests that, in part, traditional medical systems survive because they are effective. Evaluative studies, some including observations by medically trained personnel, document the extent of spiritual healing success.

Arthur Kleinman and L. H. Sung interviewed twelve patients of a Taiwanese shaman before and after they received shamanic treatment.[17] They polled patients regarding treatment outcome and satisfaction and compared these findings with their data from surveys of patients of Western-style physicians. Ten of the twelve patients (83%) reported (or their family members reported) at least partially effective treatment. Six patients (50%) regarded themselves, or were regarded by family members, as completely cured. Only two patients (17%) were listed as treatment failures. Although the researchers found no conclusive evidence that shamanic treatment was effective against a biologically based disease, the success rate as measured by client satisfaction was far higher among the shaman's patients than rates of client satisfaction among patients of Western-style doctors. About half the caseloads of the Western-style doctors are made up of people with the same kinds of disorders treated by the shaman. Among this population, not only is patient satisfaction considerably less, but also evaluations of efficacy no better.

Richard Curley monitored outcomes of healing ceremonies in Lango, Uganda.[18] He evaluated thirty-one cases of spirit possession (involving physical sickness or disability), and he determined that eighteen subjects were "cured," and three were "not cured," and one individual died. The majority of treatments in psychological cases had positive outcomes in that the spiritually negative "wind" went away.

There is a general view among health workers that traditional healers serve an important function and are effective.[19] In Ghana, for example, because cultural traditions are strong and Western-style medical facilities are limited in number, folk healers perform most primary-care services, and they are regarded by many Western-trained physicians as an important and useful part of the medical system. Many folk healers coordinate their treatment strategies with Western-style physicians, who believe that the folk healers' work often contributes to patients' recovery.[20]

Effective spiritual healers can be found in both industrialized urban and nonindustrial rural environments. Deborah Glik conducted a comparative study of participants in spiritual healing groups in Baltimore.[21] Subjects who were treated by spiritual healers were more satisfied with results than those who consulted medical practitioners.

These studies describe controlled evaluations of the efficacy of spiritual healing. A far larger collection of reports include anecdotal accounts that also indicate the success of spiritual healing.[22] Some researchers argue that the tendency for certain folk healing methods to survive in industrialized societies implies that these practices are effective.[23]

> Folk healers are given an empirical test each time someone domes to see them. If specific relief is not given within a reasonable time period, the folk healer will lose his or her practice. Thus, the laws of learning apply to folk healing as they do to other aspects of our behavior. Behaviors that reinforced continue, while those that aren't reinforced fade. The fact the traditional medicine has persisted in a culture with advanced medical technology suggests that there has been some level of efficacy in the folk treatment.[24]

The literature indicating that spiritual healing can alleviate mild psychosomatic and psychologically based disorders is so extensive that the claim that it is effective is almost axiomatic. Some observers feel that shamanic healing is *more* effective than psychotherapy. Studies indicate that African spiritual healers generate far higher "cure rates" for psychosis than do psychotherapists.[25] This difference may be because the spiritual healer is better able to manipulate factors within the community.

Research pertaining to the effectiveness of psychotherapy sheds light on the means by which spiritual healing is able to demonstrate equivalent (or superior) effectiveness. Studies comparing different types of psychotherapeutic techniques with each other and with other treatments found that all patients provided treatment tended to benefit but that there were insignificant differences in proportions of patients who improved among the treatment strategies.[26] Research comparing professional therapists to paraprofessional therapists (self-help groups, self-administered treatments, etc.) indicate similar rates of success. The variables "amount of training" and "experience of professional therapists" are not significantly correlated with client outcomes.[27] This "tie effect" (where all therapies seem equivalent) may be a result of common beneficial elements, elements that exist within indigenous treatment systems. It would seem that any method that creates a therapeutic environment, generates client expectations, and modifies client behavior can be effective. This theory coincides with psychoneuroimmunological finding that emotions affect the immune system: ritual suggestions can shape feelings and behavior, which in turn affect health.

Spiritual Healing as Treatment of Psychosomatic Maladies

Anthropologists note that a high percentage of symptoms reported by spiritual-healing clients are likely to have a psychosomatic basis, and many conclude that spiritual healing is most effective for symptoms that have psychological components. We can evaluate this hypothesis by comparing lists of symptoms that anthropologists observe to be cured by spiritual healers with symptoms verified in the hypnosis research literature as treatable. Clinical studies document that hypnosis is particularly effective for pain, asthma, warts, headache, burns, bleeding, gastrointestinal disorders, skin disorders, insomnia, allergies, psychosomatic disorders, and minor psychological problems.[28] Because hypnotic therapies can affect systems typically regarded as outside of conscious control, they may alleviate problems beyond those labeled as psychosomatic. For example, clinical studies indicate that hypnosis can change the response of human skin to heat,[29] probably through reducing edema and fluid retention following exposure to thermal injury.[30]

Hypnosis also may accelerate healing[31] due to the hypnotic control of blood flow.[32] If a patient cuts off blood flow to a cancerous tumor, for example, the growth could wither away. This theory explains some of the more anomalous healing stories collected by folklorists. One story in my North Carolina folklore collection was provided by a man whose doctor had informed him that his cancer was incurable. After receiving this prognosis, he talked with members of his church: "I asked them all to pray for me and received special blessing from some of the leaders of the church. Reverend Hunt also anointed me with spiritual oil. When I went to the doctor, he found no traces of cancer in my system. He called five other doctors from S——— Hospital to run tests on me. All the test results showed no trace of cancer." Although skeptics might argue that this story has been embellished, it is possible that the man responded hypnotically to the symbolic ministrations and subconsciously cut off the blood flow to his tumors, causing them to disappear. This explanation might even be considered probable in light of hypnosis studies.[33]

But, as stated above, the typical healings that anthropologists describe are less anomalous. Kaja Finkler observed that diarrheas, simple gynecological disorders, somatized symptoms, and mild psychiatric disorders were most amenable to the ministrations of the Mexican healing cult she observed.[34] Although herbal components may have helped reduce diarrhea and alleviate gynecological problems through nonpsychological processes, the other symptoms are amenable to psychotherapy, particularly hypnotherapy.

Kleinman provides a parallel list of disorders most treatable by spiritual healers in Taiwan: acute, self-limited (naturally remitting) diseases; non–life threatening, chronic diseases in which addressing psychosocial and cultural problems is a larger component of clinical management than biomedical treatment of the disease; and secondary somatic manifestation (somatization) of minor psychological disorders and interpersonal problems.[35] He notes that somatization is common among both indigenous and professional patient populations.

Wen-Shing Tseng observed that nearly 70 percent of the Taiwanese psychiatric outpatients in his study sample presented somatic complaints to the psychiatrist at the first visit.[36] Somatic complaints can be treated directly through suggestion; psychological and social issues can be treated indirectly through "a symbolic therapy that does not recognize them as such and, therefore, avoids stigmatizing the patient."[37]

Kleinman and Sung documented the types of problems for which a sample of shamanic clients in Taiwan sought treatment. Of 122 clients interviewed, 45 percent sought treatment for sickness, 27 percent sought determination of their fate or treatment of bad fate, 19 percent came because of business or financial problems, and 9 percent sought help for personal and family problems. Common forms of sicknesses included acute, but not severe, upper respiratory or gastrointestinal disorders. Kleinman notes that chronic disorders such as low back pain, arthritis, and chronic obstructive pulmonary disease were present, but that chronic, nonspecific complaints, called "functional disorders" or labeled "neurasthenia" by Western-style doctors, were more prevalent. He believes that these reflect somatization with depression, anxiety neurosis, hysteria, or other psychological problems.[38]

Bruce Kapferer observed that among people seeking exorcisms in Sri Lanka, stomach disorders (pains, diarrhea, constipation, vomiting, and so on) were most prevalent, followed by body fever (hot and cold flashes, high temperature, and shivering).[39] His case stories, one of which will be reviewed later, portray psychologically based symptoms related to anomalous experiences. Such problems are amenable to hypnotherapeutic treatment.

Robin Horton lists typical disorders treated by African healers: gastric and duodenal ulcers, migraines, chronic limb pains, certain kinds of paralysis, hypertension, diabetes, and dermatitis.[40] These symptoms often represent psychologically based disorders.

Vincent Crapanzano found that Moroccan Hamadsha curers were often effective when dealing with paralysis, mutism, sudden blindness, severe depression, nervous palpitations, paresthesia, and possession.[41] This supports the psychological-basis argument.

As mentioned previously, folklorists sometimes hear accounts of extremely anomalous healings. Although few provide lists of symptoms most frequently relieved by folk treatments, David Hufford notes that healers cure a wide range of problems:

> Folk healers treat practically the whole range of ills known to modern man—warts, arthritis, colic, sexual dysfunction, skin disorders, cancer, etc.—and in addition some that are unrecognized by medical science, such as *mal ochio* or evil eye, soul loss, and the effects of witchcraft.[42]

Yet the folklore literature reveals that certain symptoms appear more amenable to treatment. These include reducing pain, curing warts, stopping blood flow, healing burns, facilitating childbirth, and dealing with the psychological components associated with other physical disorders.[43] Because folklore and hypnotherapy lists are similar, we would suppose that these methods are based on common mechanisms.

The anthropology and folklore lists include symptoms one would expect to be treatable through psychological means. This conclusion coincides with Erika Bourguignon's review: she notes the prevalent observation that "success of religious healing is greatest where there is a great psychological involvement in the illness, where the disorder is primarily of a psychosomatic or hysterical nature."[44]

Skeptics often demand that alternative healing methods be evaluated using control groups and double-blind experimental procedures. Because most spiritual healing systems are based on placebo processes, this method is not always appropriate. An experimental evaluation of spiritual healing that compares the spiritual strategy to a placebo would pit two placebos against each other. With a sufficient sample size, however, the procedure stimulating greater belief and expectation would be proven effective. Hypnotherapeutic strategies gain a degree of effectiveness from the placebo effects they stimulate: methods that most increase expectations are most effective.[45]

Although it is possible that spiritual healers sometimes produce miraculous cures, the hundreds of cases I observed in Japan, China, Thailand, Sri Lanka, Korea, Taiwan, and the United States, included no healings that could not be explained in terms of hypnotic and placebo processes. Although a number of controlled experimental studies suggest that paranormal healings occur,[46] these studies are not sufficiently replicable to attract mainstream scientific attention.

Universal Features within Spiritual Healing

Observers who are familiar with the literature on hypnosis invariably perceive correspondences between spiritual healing ritual and hypnotic processes. First, spiritual healing and hypnotic processes both begin with an inductive ritual. Repetitive performances, drumming, chanting, prayer, certain postures, sensory deprivation, fasting, and other religious practices either induce hypnosis or provide cues signifying the value of hypnotic response. Second, in both processes participants respond to induction. Magico-religious altered states of consciousness have special characteristics: alternations in thinking, changes in sense of time and body image, loss of control, changes in emotional expression, perceptual distortions, changes in meaning and significance, sense of ineffability, feelings of rejuvenation, and hypersuggestivity.[47] These features coincide with hypnotic response. Third, the two processes have similar outcomes. Many of the symptoms that anthropologists observe to be treatable through spiritual healing are also treatable through hypnotherapy. And patient outcome sometimes has a temporary quality that suggests hypnotic response.

The literature linking spiritual healing to hypnosis, although ignored by many anthropologists reviewing spiritual healing, is massive. Frank Mac Hovec cites "abundant evidence which shows that hypnosis or a similar induced altered state of consciousness was used in ancient Greece, Egypt, India, China, Africa, and pre-Columbian America."[48] F. L. Marcuse collected contributions from around the world describing the use of hypnotic processes in indigenous folk traditions.[49] Observers have noted that spirit mediumship,[50] Chinese *qi gong*,[51] Filipino psychic surgery,[52] and spirit possession[53] reflect hypnotic processes. Even malevolent occult practices, such as the evil eye, have been interpreted as having a basis in hypnotic response.[54] John Schumaker links not only spiritual healing but all religious ritual with hypnotic processes.[55] Schumaker's arguments make sense to anyone who has experienced hypnosis: religious rituals have repetitive qualities that can induce trance in people who are highly hypnotizable. Schumaker's theories coincide with the evolutionary paradigm.

Anthropological observers often describe characteristics within shamanic ceremonies that reflect hypnotic processes. Hunting and gathering societies are of particular interest because their practices provide insights into possible patterns of culture during Paleolithic eras:

> According to scientists who have compared the social and economic organization in different groups of contemporary gatherer-hunters, these societies have more in common with each other than with their agricultural, pastoral, and industrial neighbors. . . . Any group of people who had to live off the land would face similar ecological problems and would probably invent a roughly similar system. It seems reasonable to suggest, then, that this pattern—or more properly, this range of patterns—prevailed in most human societies before the agricultural revolution and during much of the course of human evolution.[56]

One well-studied hunter-gatherer healing system is that of the Kalahari *!Kung*.[57] Although not all anthropologists agree, *!Kung* methods appear to be ancient, suggesting a link to Paleolithic practices. The primary *!Kung* religious ritual is a healing dance that usually lasts from dusk to dawn and occurs as often as twice a week. The band's women gather around a fire, singing songs and rhythmically clapping their hands, while the men dance in a circle. Particular men and women pass into trance and place their hands on the bodies of infirm individuals in order to heal them. This trance activity, known as *!kia*, is thought to be brought about by an energy that the *!Kung* label *n/um*.

N/um is considered to be a powerful and mysterious force that can both induce *!kia* and combat illness. The master puts *n/um* into people in order to heal them. *N/um* also gives the *!kia* master extraordinary powers such as x-ray vision and the ability to see over great distances and to handle hot coals. One master who is normally blind can see during the *!kia* state, and many masters have handled and walked on glowing hot coals without harm. Such performances seem to be related to hypnotically governed mind-body processes (a hypothesis that I will discuss later). The means of attaining *n/um* coincide with religious hypnotic inductions all over the world: rhythmic singing, dancing, clapping, and chanting. Healers describe a direct relationship between the enthusiasm with which the songs are sung and their ability at obtain *!kia*. If the singing is interrupted, the healer feels confused,[58] a response that is equivalent to a subject's sensation when hypnotic induction is interrupted. The *!Kung* perceive *n/um* as being extremely hot, an sensation that can be produced by hypnotic suggestion. My hypnotized volunteer, whom I described at the beginning of chapter 1, would have experienced an extreme heat sensation if I had suggested it to her. If it was part of her culturally shaped expectations regarding trance, she would have experience *n/um* without my mentioning it.

Unequal distribution of *!kia* ability corresponds with the unequal distribution of hypnotic talent in industrialized societies. Standardized tests for hypnotizability in Western settings indicate that about 15 percent are highly hypnotizable, a rate equivalent to the percentage of *!Kung* who become *!kia* masters. Findings regarding the psychological characteristics of *!Kung* healers coincide with current knowledge regarding hypnosis. Richard Katz measured psychological differences between healers and nonhealers among the *!Kung:*

> There are several variables which seem to increase the likelihood one would become a master of n/um. First, !kia families seem to exist—if your father has n/um, it is likely that you will get it. Also there are individual predisposing factors. For example, if you are very emotional, you are more likely to become a n/um master. Experience with intense emotions could be good preparation for the deeply emotional !kia experience. Moreover, !Kung who have a richer fantasy life, who have more access to their fantasies and are more able to accept them, are more likely to become masters of n/um. Since fantasy is an altered state of consciousness, these qualities could again be excellent preparation for contacting and accepting another altered state, that of !kia.[59]

Katz notes that it is highly unlikely that the attributes contributing to trance skill are completely innate or completely learned. Trance skills have genetic components; certain families identify themselves as "healing families," others as "families without n/um."[60] Similarly, hypnotizability is linked to fantasy-proneness and has been demonstrated to have a genetic basis.[61] Of course, some elements within hypnotizability are learned—and, in parallel fashion, the *!Kung n/um* master does not attain this role without socialization.

The process of selecting a *n/um* master illustrates the shamanic syndrome that is observed in other nontechnological cultures. Potential shamans often suffer from psychosomatic disorders and are cured of them as a part of their training.[62] They often report fantasy episodes and paranormal experiences at a young age. These characteristics are correlated with hypnotizability and the thinness of cognitive boundaries, variables that can be measured with standardized tests.[63]

People who are capable of trance performance can stimulate audiences' expectations, enhancing healing. Trance performances can include startling anomalous effects that seem to be beyond the capability of normal individuals. All over the world people in trance demonstrate feats of immunity to heat and pain. *!Kung* healers sometimes walk over a bed of hot coals or hold coals in their hands during their healing trance. Katz describes the activities of one healer:

> Kana continues to walk around in a state of kia, like a tightrope walker. He is healing people who are sitting at the little fires on the outskirts of the dance fire. He does not shriek as he pulls the sickness, but just presses his hands in the area of the chest, flutters them, and moans. He then goes back to the central dance fire, picks up several reddish-orange coals, and rubs them together in this hands, then over his chest and under his armpits. The sparks fly. He drops the coals back into the fire just as the singers begin to scatter.[64]

Such performances support folk ideologies about healing: "The Kung do not see working with the fire as extraordinary; they are not surprised that they are not burned. Their explanation is simple: when the num in their body is boiling and as hot as the fire, they cannot be burned when working with the fire. When their num is dormant or cold or cooled down, they can be burned."[65]

!Kung healers, like shamans all over the world, are also thought to demonstrate extrasensory perception. Even though academics often disparage such reports, a large percentage of people in all societies (in the United States, over 50%) report extrasensory experiences, which lead many to believe in magical abilities. Parapsychological evidence suggests that trance facilitates extrasensory performances.[66] Charles Honorton's review of eighty-seven experimental studies indicates that meditation, hypnosis, induced relaxation, and sensory deprivation significantly improved performance on tests of paranormal abilities.[67] A theory explaining this relationship suggests that extrasensory information is generally blocked by the "noise" of normal consciousness but that some altered states of consciousness reduce this noise, allowing extrasensory information to be received. Even if paranormal phenomena do not exist, many audiences believe that they have witnessed extrasensory performances and these perceptions support their belief in spiritual healing.

Many anthropologists believe that they have seen shamans gain information paranormally. Stanley Krippner cites ethnographic accounts of ostensible psi phenomena from field studies with West African diviners, Jamaican diviners, Zulu shamans, and a Haitian voodoo priest.[68] Joseph Long provides over a hundred citations of possible paranormal phenomena noted by anthropologists.[69] Referring to the ethnographic literature as evidence, Mircea Eliade argues that the authenticity of these phenomena is indisputable.[70] It would be more precise to argue that the nature and prevalence of paranormal phenomena are sufficient to induce belief among many indigenous observers and among scholars such as Krippner, Long, and Eliade. Belief generated by performance is important because it results in hypnotic and placebo effects.

Although Katz does not evaluate the therapeutic effectiveness of *!kia* dances, the *!Kung* perceive that their dances are beneficial.[71] Many of the medical problems that the *!Kung* experience (physical trauma, complications of childbirth, infertility, burns, mental disorders) could be alleviated through hypnosis and apparently are successfully treated by healers who use hypnotic processes.

Observers who are familiar with hypnotic effects note a direct relationship between spiritual healing and hypnotic processes. For example, Wonsik Kim views Korean shamanism as a form of hypnosis for a number of reasons. The repetitive nature of shamanic praying, singing, ringing of bells, and beating of drums induces trance and provides inferred suggestions. The client appears to be in trance: facial expression, words, and movements indicate a hypnotic state. The client moves an "enchanted" stick in response to questions, and the unconscious response aids diagnosis in a manner similar to that which occurs in hypnotherapy. The shaman's performance includes special monotonous intonations of speech that are synchronized with the beating drum, contributing to trance induction, and shamanic practitioners experience difficulties in inducing trance in some clients just as hypnotherapists do.[72] Equivalent percentages of highly responsive and nonresponsive individuals are observed among both shamanic clients and experimental hypnotic subjects. Kim concludes that "there is no difference between the two phenomena, only in the ways of studying and measuring them. . . . the two are just different ways of study and utilization of the same truth."[73]

Shamanic Performance and Hypnosis: Participant Observation

My participant observation supports Kim's conclusions. Shamanic healing systems involve performances in which the practitioner goes into trance and engages in actions that stimulate belief. These performances link hypnotic processes with psychosomatic healing. Performances often involve several elements: *spiritual flight,* in which the shaman mentally travels to a distant location; demonstrations of alleged extrasensory perception; ostensible communication with deceased people; demonstrations of psychokinesis (apparently accomplished with sleight-of-hand magic); and heat-immunity feats, in which the shaman handles extremely hot objects or walks over burning coals.[74] These strategies stimulate expectancy, which generate hypnotic and placebo results. Hypnotizable observers need not be in trance for suggestion to be effective (see chapter 4).

I became convinced of the relationship between hypnotic and shamanic performance when I was a participant observer of firewalking in Japan, Sri Lanka, and the United States. Performers who defy heat injury are thought to be able to accomplish other magical feats, particularly healing. People who witness heat-immunity feats often increase their belief in the performer's ideology, and placebo effects are enhanced.

Shamanic healers throughout the world demonstrate heat-immunity feats: modern exhibitions occur in India, China, Japan, Singapore, Polynesia, Sri Lanka, Greece, Bulgaria, and other localities. Stories of heat-immunity feats are part of the folklore in many societies. Stith Thompson's *Motif-Index of Folk-Literature* includes many numbered categories and subcategories related to heat immunity (for example, D1841.3, "Burning magically evaded" and D1841.3.2.3, "Red hot iron carried with bare hands without harm to saint").[75] Each listing includes citations of stories from different cultures. Because fire tending has been a part of the human experience for over 500,000 years, it is likely that heat-immunity feats were part of Paleolithic shamanic performances.

I first attempted firewalking among the Shingon Buddhists at Mt. Takao, Japan, in March 1983. Japanese Shingon Buddhism has a long history of esoteric, mystical performance derived from Chinese traditions. The Buddhist priests conduct a dramatic ceremony that includes colorful costumes, lengthy chanting, and the shooting of arrows to drive away evil spirits. Believers write their ailments on *nadegi* (literally, "rubbing boards"), which they hold against infirm parts of their bodies. *Nadegi* are then used as firewood for a huge fire, which is allowed to burn down to a twenty-foot coal bed. The priests then walk across the hot coals and allow lay people to follow them.

I decided that the best way to understand firewalking was to do it myself. I joined the throng of lay people who followed the priests. The priests had trampled out a pathway of hot ash through the bed of glowing coals. I felt relieved that I would not be required to step on a glowing coal. But as I waited behind an old woman, a priest raked the coals even, and the ash pathway disappeared; it was covered by coals. I was afraid. I followed the old woman onto the coals as she walked at a terrifyingly slow pace. My mind was blank with fear and excitement. I was stepping on red hot coals, yet not being burned! After I completed my walk, my feet tingled but were not harmed.

Traditional peoples attribute firewalking success to religious forces. Western observers, particularly those associated with a skeptical movement that debunks paranormal claims, provide alternative explanations. Heat resistance can be explained as a product of the Leidenfrost effect, the propensity for water droplets on a hot skillet to dance around on a vapor cushion so that they remain in liquid form longer than usual in spite of the high temperature. Mayne Reid Coe Jr. used this theory to explain his ability to hold red-hot iron bars in his hands. He assumed that moisture from his hands vaporized rapidly due to heat from the bars and that the vapor provided a thermal cushion, preventing his skin from burning.[76] Jearl Walker accepted this theory as valid. He demonstrated the Leidenfrost effect experimentally and assumed that the process explained his own firewalking feat.[77]

Bernard Leikind and William McCarthy attributed the failure of human skin to burn during firewalking to low thermal conductivity between the coal bed and the firewalkers' feet.[78] Low thermal conductivity is illustrated when a person puts his hand in a hot oven. Although the oven air is extremely hot, the heat is not conducted rapidly from air to hand. Although the low-thermal-conductivity theory does not explain Coe's handling of red-hot iron bars, it seems to be a satisfactory explanation for many firewalking performances. Most firewalkers walk relatively rapidly. Their feet are in contact with the hot surface for about half a second, and relatively little heat is transferred to the bottom of the their feet during that brief period of time. I assumed that the low-thermal-conductivity theory was the best explanation for my firewalk. The Leidenfrost effect requires relatively high temperatures, and the fire beds at Mt. Takao were continually cooling as people walked over the coals.

Some heat-immunity feats seem to be anomalous. My literature review uncovered a later report by Coe in which he described his discovery that he was unable to respond to verbal questions while holding the red-hot iron bars. He decided that trance contributed to his ability to hold red-hot iron bars and that all his feats had required trance, even though he previously had not realized it.[79]

Anthropologist Loring Danforth links firewalking and spiritual healing. During his observation of the firewalking Anasternaria of Greece, he observed many of the features within the shamanic complex. People had been suffering from medical disorders that were resolved when they joined the Anasternaria firewalking cult.[80] As in the case of the Shingon Buddhists of Japan, the Greek cult members' demonstrations of heat immunity supported beliefs that contributed to spiritual healing.

In July 1993 I took part in the Esala Festival at Kataragama, Sri Lanka. Tamal Hindu pilgrims engage in masochistic rituals to demonstrate the power of their faith. Some pierce their tongues, cheeks, backs, and arms, and roll on the ground for great distances. Entranced people put hooks through their backs and hang from scaffolds. It seems likely that these pain-denial feats entail hypnotic processes because most involve trance and performers claim to feel no pain. The high point of the Esala Festival demonstrations is a firewalking ceremony before a large crowd early in the morning.

I joined the week-long ritual preparations of those who elected to firewalk. The firewalkers vow to forgo eating meat, telling lies, drinking alcohol, and engaging in sex for one week prior to the event in order to gain divine protection. I interviewed holy men who had firewalked in past years. They told me that firewalking was a mental feat, that "forgetting that fire can burn" allows one to walk safely. When I described the low-thermal-conductivity theory, they laughed. "If a normal person touches a hot coal," one told me, "he will be burned. All children learn this. To walk safely, forget what you have learned."

About fifty Tamal Hindus and Singhalese Buddhists took part in the ritual preparations. My fellow pilgrims offered a variety of explanations for how the feat could be accomplished. Most were fulfilling vows they had made to a Hindu deity, Murugan, Shiva's son, and believed that fulfilling the vows was instrumental for success. Others advanced explanations derived from their Buddhist faith. A few participants, such as the holy men, attributed success to cognitive powers. Without realizing it, I violated a dietary prohibition by eating a hard-boiled egg a few days before the ceremony (my Singhalese language skills did not allow me to learn that eating eggs was a violation of the vegetarian vow). I decided to continue my firewalking preparation, placing faith in the low-thermal-conductivity theory.

I observed that many of my fellow participants were true hypnotic performers. One passed into trance while the group engaged in a ritual dance. He fell to the ground, shouting. Others sat motionless for long periods of time, meditating. If firewalking was a test of faith, I could not be sure of success. I told myself that I was a Westerner who trusted the laws of physics. I decided to walk, though quickly, because I believed that low thermal conductivity would protect me from being burned. At the same time, I hoped that the religious atmosphere would aid me if faith was required.

On the evening before the event, we firewalkers participated in a dancing ritual, a form of hypnotic induction. Like many in our group, I was afraid. I had been told of a woman who had burned to death after falling on the hot coals a few years before. I was warned that all those who leaped sideways off the coal bed were seriously burned.

Early in the morning, after many hours of ritual dancing, we walked, one by one, across a twenty-foot-long, red-hot bed of coals. My first step was extremely painful. I walked as rapidly as possible, talking long steps. The crowd applauded wildly. I was the only Westerner to have firewalked safely at Kataragama. After we walked to the nearby Hindu temples to receive blessings, the Buddhists invited me to accompany them to the distant Buddhist stupa for further rituals. People called me "Swami" because I was "one of those who had done it." I felt a great sense of relief and exhilaration.

More than an hour after my "success," I discovered that my feet were beginning to blister. I was extremely disappointed. I went to the first-aid station and was surprised to find that some of the most religious firewalkers were already there. One man who had spent much time teaching me the Hindu religious shouts continued to shout these exclamations as he waited for treatment.

I looked at the station register and observed that over a third of those who had walked had visited the first-aid center for treatment. The crowds had gone home, thinking that no one had been injured. The Hindu medic suggested that I must have violated one of the prohibitions or else I would not have been injured (ironically, he was correct).

It appeared to me that firewalking success was correlated not with public displays of religiosity but with hypnotic talent. Some of the most vocal Hindus had been burned while those who had demonstrated trance behaviors had walked successfully. A young boy carrying a coconut had shown particular skill by walking slowly and gracefully across the coals. I had no way of evaluating which firewalkers had violated prohibitions although I later interviewed one injured man who claimed to have kept all his vows. Believers explained to me that firewalkers such as myself would have been more severely injured if we had not been protected by spiritual forces.

My blisters required about three weeks to heal. While I recovered, a master firewalker explained that I had been injured because I had hurried rapidly over the coals. My pattern of blisters suggested that I had pressed my toes and feet too heavily into the fire bed in order to propel myself forward. He believed that it I had walked evenly, like the boy, my feet would not have dug so deeply into the coals, and I might have passed over them safely. Firewalking can be viewed as a gymnastic feat in which walking evenly and lightly contributes to success.[81]

During this period, Sri Lanka was consumed by ethnic rioting and strife.

Many Tamal shops had been burned and my personal injuries seemed minor compared to the suffering of others. Some fires raged out of control and entire villages were destroyed. The turmoil forced me to consider the role that religious beliefs play in both causing and alleviating suffering. Obviously, religious beliefs are not always functional because they stimulate sectarian conflicts that sometimes destroy whole societies.[82]

I suspect that some people use hypnotic processes to cope with stressful situations. People with hypnotic talent are able to ignore pain. They can firewalk casually like the young boy. They may also use their self-hypnotic talent to cope with psychological stress. People who are able to ignore unresolvable problems by dissociating may maintain psychological stability.

Stephen Kane argues that the mind can affect physiological processes, protecting the firewalkers.[83] He proposes that inferred suggestions can cause the body to mobilize peripheral vasoconstriction in the area exposed to high temperatures. Although skin cells have been exposed to high temperatures, blisters do not form and the firewalker suffers no injury. Robert Sammons summarizes the arguments describing how hypnosis can heal burns, describing processes that would benefit firewalkers exposed to heat trauma:

> Hypnosis appears to effect the healing of burns by three separate measures. First, there is immediate relief to the patient by the reduction of the pain. Second, hypnosis reduces the extension of the injury by reducing the initial inflammatory response. Later hypnosis promotes healing by increased blood flow, which stabilized area metabolism and carries away waste products and damaged tissue.[84]

Sammons's arguments are supported by clinical research. Hypnosis has been found to be effective for the treatment of burns, particularly within the first two hours after burning when it can limit inflammatory reaction to thermal injury.[85] Well-controlled studies indicate that hypnosis reduces burn pain better than placebos.[86] A firewalker who experienced no pain or inflammation, and whose body quickly repaired damaged areas, would walk "safely" even if surface skin cells were destroyed.

I formed the hypothesis that the degree of firewalking success is positively correlated with the degree of hypnotizability. This hypothesis coincides with experimental studies indicating that hypnotic suggestion can repress or increase inflammatory response and blood flow. Suggestions can also cause a hypnotized person who is touched by a piece of metal that is not hot to blister as if burned.[87] This suggestion theory is also supported by evidence derived from surveys of firewalking workshop participants in the United States. A number of studies indicate that cognitive processes affect firewalking success. Julienne Blake polled firewalkers regarding their beliefs and had medical personnel examine their feet after a firewalk. Those who had relied on external factors (faith in the workshop leader, for example) had significantly less incidence of heat trauma and blistering than did those seeking protection from internal factors.[88] Ronald Pekala and B. Ersek administered the Phenomenology of Consciousness Inventory (PCI) and the Dimensions of Attention Questionnaire (DAQ) to twenty-seven firewalkers. Scores on these scales showed a difference between the participants who had been burned and those who had not.[89] J. A. Hillig and J. Holroyd "extended and partially supported the investigation reported by Pekala and Ersek."[90] They also found that "an alteration in attention may be more important to successful firewalking than an alteration in consciousness."[91]

Firewalking might be regarded as a test of the ability to focus attention. It is probably a better measure of the mind's capacity to affect the body than standard laboratory hypnotizability tests. It also provides a metaphor associated with therapy. The fear of being burned while walking across a coal bed is parallel to other fears. Fear can be dysfunctional, inhibiting a person's performance. The stress of walking across a coal bed is a metaphor for the stresses of life that reduce immune-system efficiency. People who respond in a more relaxed manner as a result of ritual suggestion have a survival advantage.

While I was recovering from the injuries I had suffered in Sri Lanka, I began an intense meditation and self-hypnosis program. I armed myself with faith in hypnosis. Skeptics argue that hypnosis cannot be adequately defined, but people who are hypnotized (or who hypnotize themselves) believe that they experience an altered state of consciousness. It is logical to assume that suggestions provided under hypnotic circumstances will have effects, at least expectancy effects. In trance one feels in greater contact with the subconscious mind.

Thus far I have firewalked successfully almost two dozen times. On various occasions, I led others across the coals. I assume that my hypnotic inductions and suggestions helped them to walk safely. On eleven occasions, I measured the fire bed temperature and placed heat-monitoring labels on the bottoms of my feet before walking. These measurements provide insights regarding firewalking processes. For example, in August 2000, I placed my feet at the exact places on the coal bed where I had previously measured the temperature as 963 degrees Fahrenheit (left foot) and 995 degrees Fahrenheit (right foot). The temperatures registered by the monitors on the bottoms of my feet varied (the monitors become darkened when exposed to a designated temperature). On the right foot, for example, the 190, 280, and 350 degree monitors were darkened. Yet all the 200 to 270 degree monitors, the 290 to 340 degree monitors, and all those above 350 degrees were not darkened. All other measured firewalks produced equivalent, but similarly sporadic, readings. Overall, the maximum darkened monitor temperature on each foot correlated positively with the fire bed surface temperature.

These measurements demonstrate that the bottom of my feet were exposed to temperatures above the boiling point for water. The readings also indicate a low rate of thermal conduction and an uneven transfer of heat.[92]

All of the firewalks caused a tingling sensation on both feet but, after Sri Lanka, no inflammation or blistering. The tingling sensation was increasingly uncomfortable with higher temperatures. Because the temperatures of the bottoms of my feet and of the coal bed were higher than the boiling point of water, we might hypothesize that the bottom of my feet did sustain cellular damage, but that physiological processes associated with suggestion reduced the blistering that typically follows such damage.

Firewalking is one of various ways in which shamanic performers demonstrate trance ability. More frequently, they seek to convince clients of their extrasensory ability, particularly by practicing divination. Typically an individual or group comes to the shaman with a problem. The shaman goes into trance and uses a particular method, such as consulting an oracle. This method requires the shaman to interpret his or her internal cognitions. The client believes that the shaman has gained information supernaturally, and the shaman proceeds to provide counseling.

I have many accounts that fit this pattern in my folklore collection from northeastern North Carolina (see chapters 5 and 6). People all over the world visit psychic practitioners, and many believe that the psychics' predictions are valid. One woman states:

> Everything she [the psychic] told me has come true. She told me I would get married within a year to a different man than I was going with. I didn't think that was possible, but in less than two months, I learned that my boyfriend had been cheating on me, and we broke up. The more surprising thing was that she told me I would have a baby girl within the next few years. As it turned out, I met Tom only a month after breaking up, and in a whirlwind we were married. I then found out I was pregnant, and sure enough Shannon was born within the same year. She had not told me exactly who Tom would be, but she had told me he would have lighter color hair.

Although I have never obtained information paranormally from a psychic practitioner, many parapsychologists believe that their experiments prove the existence of psychic phenomena.[93] Skeptics are not convinced for a number of reasons. First, it is possible to design more stringently controlled experiments, and skeptics contend that paranormal claims are so exceptional that greater controls are required. Second, not every study of psychic phenomena yields statistically significant results. Critics feel that the percentage of studies achieving statistical significance is not adequate, given the exceptional nature of the claim. Third, because parapsychological researchers could have been cheated or deluded, critics argue that greater levels of security are required. Various parapsychologists disagree with the critics' arguments, claiming that current experimental controls, levels of replicability, and precautions against fraud are adequate. I do not argue that the parapsychologists' claims are valid or invalid, but I do assert that shamanic attempts to demonstrate extrasensory phenomena often lead to belief within indigenous audiences and among some anthropologists and parapsychologists. Because shamans are able to inspire belief, their healing rituals are made more effective.

The Anthropological Literature

Spiritual healing is regarded as universal, existing in all societies.[94] Major anthropological reviews attempt to describe the specific features that make spiritual healing work. Both Bourguignon and Eliade note that spiritual healing is based, in part, on altered states of consciousness and suggestion,[95] a conclusion that infers hypnotic processes. Although other reviews are more abstract, I argue that the incidence of hypnotic processes within all spiritual healing systems creates recurring themes within the anthropological literature.

John Comaroff argues that universal features include the provision of codes for reordering the disrupted relationship between patients' physical and social states and for rendering sensible their apparently chaotic experiences.[96] Like other reviewers, he describes no mechanisms for how this relationship is reordered or how these experiences are made sensible, but we should note that hypnotic processes could accomplish these objectives.

Kleinman describes three stages of indigenous healing rituals: "The sickness is labeled with an appropriate and sanctioned cultural category. The label is ritually manipulated (culturally transformed). Finally a new label (cured, evil) is applied and sanctioned as a meaningful symbolic form that may be independent of behavioral or social change."[97] Kleinman provides a complex model that does not include mechanisms, but again hypnotic processes could generate the transformations he points out.

James Dow describes four universal features in ritual healing systems. First, experiences of healers and the healed are generalized with culture-specific symbols in cultural myth. Second, suffering patients come to the healer, who persuades them that their problem can be defined in terms of the myth. Third, the healer attaches the patient's emotions to transactional symbols. Fourth, the healer manipulates the transactional symbols to help patients transact their own emotions.[98] As with the previous theories, these processes could be accomplished through hypnotic suggestion.

These formulations have been criticized for failing to specify how spiritual healing actually works.[99] Spiritual healing entails more than labeling sicknesses, manipulating labels, and applying new labels. None of these theorists stipulate how persuasion occurs or how healers create a disposition to be healed. The Japanese Shingon firewalking ceremony described previously includes symbolic activity (rubbing boards on infirm parts of the body and then burning the boards), but the persuasive power of the ceremony involves heat-immunity feats. Saying that ritual activities manipulate labels and apply new labels misses the need for the emotional impact associated with ritual performance. Magical feats capture people's attention so that they subconsciously accept therapeutic suggestions.

The anthropological models should be reformulated to include hypnotic response. In terms parallel to those in Dow's text, I argue that successful ritual performances bring about an increase in expectation that allows the healer to persuade others that problems can be defined in terms of a myth. Hypnotic mechanisms permit the healer to connect people's emotions with transactional symbols so that manipulation of these symbols provides hypnotic suggestions. This often requires adjustments of role relationships and assignment of new roles, processes facilitated by suggestion. The previous theories are not false; they are merely incomplete. Manipulation of symbols provides waking suggestions for hypnotizable people so that hypnotic induction is not always required.

Unless researchers administer tests for hypnotizability, their field observations cannot prove that spiritual healing involves hypnotic processes.[100] The recurring features surrounding spiritual healing merely *imply* that these processes occur. I hypothesize that shamanic rituals constitute hypnotic inductions, that shamanic performances provide suggestions, that client responses are equivalent to responses produced by hypnosis, and that responses to shamanic treatment are correlated with patient hypnotizability. These testable hypotheses are more amenable to empirical evaluation than previous anthropological formulations. Although present standardized tests of hypnotizability are culturally specific, hypnotizability could be operationalized in a manner appropriate for any society. Questionnaires measuring the permeability of cognitive boundaries and the frequency of psychic, dissociative, and absorptive experiences should also predict a person's capacity to be spiritually healed.

Shamanic Healers: Participant Observation

Between 1982 and 1986 I observed spiritual healers in Japan, Korea, the Philippines, Taiwan, the People's Republic of China, Sri Lanka, and Thailand. I interviewed dozens of healers, observed their treatment methods, and talked with their clients.

One Taoist healer, Mr. Li, was particularly instrumental in causing me to think about the common features among ritual practices. "Would a ritual within one society work for people from another culture?" he asked.

"It doesn't seem likely," I replied. "The Christians pray to Jesus, but they think that belief in Jesus allows the healing. Only rarely are Muslims, Jews, Hindus, or Buddhists healed by Christian rituals." (At the time I had not considered the implications of the Muslim man's cure by the Christian psychic surgeon, which I described in the introduction.)

"I believe that the spirits who help me would also help foreigners. Their belief is not important. We could do an experiment to see," Mr. Li offered.

In order to test his hypothesis, Mr. Li asked that I assemble eight Westerners who were unfamiliar with Taoist rituals. He planned to conduct a "visit to heaven and hell" ceremony to determine if he could transport Western participants to the same Taoist heavens and hells that his Taiwanese clients visited. The ceremony is sometimes performed to help grieving families, but I did wonder what people believed had happened during the ritual. When asked, one previous participant said, "We visited heaven and hell. The ceremony is like a bus; it takes you to those places."

I found eight non-Chinese who agreed to participate and were unfamiliar with the Taoist ceremony. Mr. Li began the ritual by invoking local gods to provide protection and assistance. He burned (and had us burn) great quantities of religious paper money (special currency used to pay the deities for attending the ceremony). It was clear from the participants' comments that they were not believers. One asked if his skepticism would keep the ceremony from working. None had heard of the local god whom Mr. Li considered his main source of inspiration, and none had previously visited heaven or hell.

We sat on benches, and Mr. Li blindfolded us while his assistants chanted and played percussion instruments rhythmically. I quickly recognized the hypnotic quality of the ritual. Mr. Li repeatedly stated that we would feel progressively lighter as the ritual continued and that we would eventually rise up from the bench. I *did* feel lighter but decided to resist his suggestions so that I could later become an observer. After much chanting, drumming, and verbal suggestion, a woman stood. She was led to the back of the temple and became the focus of attention. The other participants and I were allowed to remove our blindfolds and watch her.

Mr. Li led the woman through a series of mental exercises based on religious images. At the shaman's suggestion, she traveled in her mind to hell and later to heaven, and she talked with the people she saw there. As predicted by Mr. Li, she perceived images of heaven and hell that were parallel to the images held by Taiwanese people. The performance was a kind of morality play: those in hell were suffering due to their wrongful behavior while residents in heaven enjoyed the fruits of their virtuous lives. From my vantage point it seemed that the woman was responding to hypnotic cues, and like most hypnotized people, she creatively constructed images coinciding with suggested themes. The notion of punishment and reward in the afterlife exists in both Taoism and Christianity.

Later I analyzed medieval European, Chinese, and Japanese accounts of near-death experiences (NDEs). Common features within these stories seemingly brought about a convergence in images regarding heaven and hell in both medieval Europe and Asia.[101] Although some elements within NDEs reflect the subjects' host societies, the human mind may be wired to perceive or construct common notions regarding life after death no matter what culture the subject is raised in.

I found it easy to interpret Mr. Li's ritual within the context of what we know about hypnosis. The preliminary part of the ceremony allowed Mr. Li to identify a hypnotizable individual who would be a good performer for the journey to heaven and hell. The woman identified herself by following Mr. Li's suggestion to stand up. She continued to respond to his verbal cues, providing the audience with a dramatic account of her visits to heaven and hell. My pre-linguistic hominid campfire ritual was derived from what I learned by watching Mr. Li's ritual. Highly hypnotizable people respond to repetitive movements and sounds. When in such environments, they go into trance and construct dreamlike stories regarding inner images. They sometimes describe these images as "more real than normal reality."

When I talked with the woman later, she labeled her state of mind as "a kind of trance" in which she was aware of her surroundings. Hypnotized people often use such terms. I later tested her and found that she was highly hypnotizable. She reported that her hypnotic experience was equivalent to the "visit to heaven and hell." Her hypnotic state apparently aided her in creatively generating images that were particularly real for her.

Mr. Li's ritual illustrates shamanic spiritual flight performance. People all over the world are fascinated by messages obtained in trance. Because not everyone can achieve this state, audiences depend on an expert for spiritual information.

We can interpret Mr. Li's ceremony within the context of Kleinman's theory. Although it is correct to say that Mr. Li ritually manipulated symbols (he burned paper money) and that he sanctioned rituals that labeled certain behaviors good and bad in a way that supported therapeutic belief,[102] it is also correct to view the ceremony as a hypnotic induction. The ritual works because some people are hypnotizable. Ritual manipulation of symbols has no value for people who lack this trait.

It is possible to test this hypnosis hypothesis. Since observing Mr. Li's ritual I have conducted my own versions of his ceremony for many different audiences in college classes, professional seminars, and academic conferences. As with my fireside ritual, I find it easy to identify highly hypnotizable volunteers and to get these people to demonstrate shamanic skills. People describe their visits to heaven and the activities of relatives at distant locations, and they even accurately predict future events. Audiences are highly entertained by these performances and often wonder if supernatural forces are involved. People who play shamanic roles are particularly affected and sometimes generate incidents they perceive as paranormal. One volunteer who was hypnotized writes:

> On the way up to heaven, I saw many clouds. It seemed to get brighter the higher I went. At the gate, I presented my pass. A hand emerged from the clouds and led me in. While inside, I saw many people socializing, walking, and just hanging out. I kept looking. As a crowd of people moved from my view, I saw a little boy. He was playing with what looked like a car. The little boy was my brother. I never knew my brother for he was in a car accident at the age of two. I told my mother of this experience. She looked at me with tear-filled eyes and said "Rhonda, he had a little crazy car he used to play with

all the time." I only know about my brother from newspaper obituaries and photos. Mom has never opened up about him and I had no prior knowledge of the crazy car. How did I see it, if it really didn't happen? I leave that up to you. I do know, however, that anything is possible when you are dealing with the human mind.

Other subjects report spirit journeys in which they see unusual events occurring at their homes that are later verified as true (a street construction project in front of the house, for example). Some of my subjects have accurately predicted winners of sporting events; surprisingly, not one has failed at this task so far.

Various observers report similar occasions where experimental subjects in trance provide information gained paranormally. Mesmerists in the 1800s, who practiced a form of hypnotism before the term was coined, found that entranced people had clairvoyant abilities (this will be discussed more fully in chapter 4). The psychiatrist Robert Bergman witnessed a similar phenomenon. He hypnotized a subject to demonstrate trance phenomena for a group of Native American medicine men. They immediately recognized the phenomenon as something they used as a diagnostic procedure—a way to gain information regarding important questions. They insisted that Bergman seek information regarding the severe drought that was thought to continue for another year. Bergman reports the results:

> When my subject was in a deep trance, I instructed her to visualize the weather for the next six months. She predicted light rain within the week, followed by a dry spell of several months and finally by a good rainy season in late summer. I make no claim other than the truthful reporting of facts: She was precisely correct.[103]

My purpose in describing Bergman's and my experiences is not to encourage belief in paranormal phenomena but to illustrate how hypnotic rituals induce belief.[104]

The anomalous claims that hypnotic rituals generate should not be regarded as *scientific* evidence supporting belief in the paranormal. Many accounts could be explained as due to the practitioner's sleight of hand and misdirection and by the observers' selective memory. Nevertheless, the hypnotic processes associated with shamanism probably stimulated anomalous perceptions during Paleolithic eras, shaping the development of religious traditions. Accounts from modern indigenous observers of shamanism support this argument. For example, John Putman, writing for *National Geographic,* interviewed a sixty-five-year-old resident of Diomede, an island close to the Bering Strait:

> The Eskimos, in their isolated villages, offer insights into the everyday life of their Paleolithic forebears. . . . Moses [Milligrock] had boyhood memories of

shamanism. "There was a man who sang and beat the drum and put a long knife through his body and pulled it out again, and he was unhurt. There was a man with a bowl, and he put water in it and you could see all over, like it was a radar. Lots of people lost in those days—and the witchcraft people saw in the bowl where the people were lost."[105]

While anthropologists who are aware of my research sometimes tell me of such incidents, they rarely mention them in their formal publications. Although it is beyond the scope of the ritual healing theory to explain extremely anomalous events, I have found such stories in many localities; they are even described in medieval Chinese documents.[106] We might speculate that these folkloric accounts are related to hypnotic processes because they seem similar to spontaneously occurring anomalous perceptions. Such traditions provide the foundation for hypnotic and placebo healing because they instill belief in shamans' magical abilities.

Anthropological Case Studies

Although anthropologists rarely observe anomalous events within the context of ritual performance, their ethnographies often reveal elements that imply hypnotic processes. Kleinman and Sung, for example, describe a case illustrating how hypnotic processes contribute to transformation and healing. Mr. Chen, a lower-middle-class Taiwanese woodworker suffered from a severe sensation of discomfort in his chest. Other symptoms included nervousness, worry, weakness, and malaise. Western-style doctors diagnosed his ailment as neurasthenia but were unable to relieve his problem. Psychiatric evaluation revealed no evidence of psychosis but demonstrated "acute exacerbation of a chronic anxiety neurosis owing to psychosocial stress, principally involving financial and business concerns." Mr. Chen's complaints "involved somatization as a secondary manifestation of his primary psychological problem."[107] Mr. Chen sought advice from Chinese medical practitioners, who proved unhelpful.

When Mr. Chen consulted a Taiwanese shaman, a *tang-ki*, the healer told him that he should rest quietly in the shrine and that the god would come to him. He was given charms and sacred ash that had been specially prepared for him. Members of the cult told him of their own cures and stated confidently that he too would be healed. Most of Mr. Chen's therapy was provided by the milieu rather than by the *tang-ki*. Mr. Chen was advised to relax, close his eyes, and try to achieve a trance state. During his third visit, he successfully became entranced, flung himself violently about the shrine, and eventually collapsed on the floor exhausted. Afterward he reported feeling completely well. During a follow-up interview two days later, Mr. Chen reported that he had been cured in a manner he considered strange. About three weeks later, he admitted that he occasionally suffered from a discomforting feeling. During interviews three and seven months later, he did not appear anxious, and he denied having experienced further acute anxiety states. Mr. Chen had become an active member of the *tang-ki's* cult.

Although Mr. Chen's cure involved special procedures and the manipulation of symbols (charms and ash), features anthropologists note to be universal within spiritual healing, Mr. Chen's healing was dependent on his ability to go into trance. This altered mental state apparently allowed him to reframe his relationship to the stressors in his life that had previously caused him such psychological and physical discomfort.

Kapferer describes parallel features within Sri Lankan exorcisms.[108] Although he argues that exorcisms work by rebuilding traditional social hierarchies, and although he seems to be unaware of the hypnosis literature, his accounts reveal the importance of hypnotic processes in spiritual healing. In one case an eight-year-old girl, Asoka, heard a hallucinatory female voice and awoke screaming. She suffered from a headache and severe stomach cramps. Because this event occurred almost a year to the day after her aunt Selfina's death, the aunt's ghost was regarded as a likely cause for the incident. Selfina had died without bearing children, a situation supporting her candidacy as a causal agent. The demon Mahasona was also regarded as a possible contributing factor.

Asoka was taken to a hospital and examined by doctors trained in Western medicine. They suspected meningitis and ordered a spinal tap. The result of this and other tests proved negative, and Asoka was released after two days.

Asoka's family sought advice from spiritual practitioners, and two exorcists concurred regarding the spiritual roots of Asoka's problem. One performed a ritual treatment thought to be powerful for three days. After the three days had expired, Asoka again heard voices and suffered a headache and stomach cramps. Because of the family's skepticism, Asoka was again admitted to the hospital, where she underwent further tests and a skull x-ray. The results were still negative.

Evidence of the need for an exorcism continued to accumulate. During the performance of a Mahasona exorcism at a neighbor's house, Asoka leaped out of bed and danced, entranced, to the site. Her action occurred at the height of the ceremony, when the malevolent Mahasona was understood to be present. After her trance ended, Asoka revealed that she had heard Selfina calling her. A senior family member suggested that the problem involved the entire family because Selfina's death could have been the result of sorcery. This argument supported the eventual decision to pay for an exorcism.

Asoka recovered, apparently as a result of this exorcism. She experienced no more headaches, stomach problems, or auditory hallucinations. Although Kapferer's study portrays the exorcism as functional for the family, Asoka's hypnotizability seems to have been the major element governing her cure. Her symptoms appear to have been psychosomatic. She demonstrated her suggestibility by hearing hallucinated voices, by immediately relapsing after the first ritual's power had expired, and by dancing in trance during the neighbor's Mahasona exorcism. The success of the final exorcism supports this argument.

My argument that spiritual healing is hypnotically based fails to capture the complex social-psychological dynamics associated with many cases. Spiritual healing often has incremental characteristics portrayed more fully by others.[109] Thomas Csordas, for example, places emphasis on people's personal experiences, and therefore his analysis provides insights into the complexity surrounding successful treatments.[110] Spiritual healing is not merely manipulation of symbols; rather, it entails an ongoing process of suggestion, acceptance of suggestion, and transformation pertaining to new social relationships and self-identity. Events originally labeled as transforming may not be permanently integrated into the person's life, and therefore they may have little therapeutic impact.[111] In chapter 6 I will describe two of Csordas's case studies, examples of group processes associated with anomalous experience.

Conclusion

Anthropological studies support three central hypotheses: First, that spiritual healing can be effective. Anthropologists have interviewed and examined patients before and after their treatment by shamanic healers, noting patient satisfaction and evaluating improvement. Their conclusions could be replicated by future studies. Second, that spiritual healing is most effective for treating ailments with psychosomatic components. This hypothesis is also amenable to further analysis. Anthropologists have documented symptoms that are most often cured by spiritual healers. The listed symptoms are similar to symptoms that are treatable through hypnotherapy. Third, that the effectiveness of spiritual healing is due in part to hypnotic and placebo processes. Shamanic healers provide autobiographies that suggest that they are highly hypnotizable; they use methods equivalent to hypnotic inductions, and successful healings often reveal elements suggesting hypnotic processes.

The ritual healing theory provides testable hypotheses leading to practical applications. Researchers could determine which people are more amenable to ritual healing. Armed with this knowledge, they could improve the efficiency of medical systems by directing highly hypnotizable people toward spiritual and psychological practitioners who could best meet their needs. Ritual or spiritual treatments would be an effective adjunct to standard medical care.

4

Wondrous Healing, Hypnotizability, and Folklore

Medical doctor Lewis Mehl-Madrona described his experiences during a healing and singing ceremony conducted by a Native American shaman named Paul:

> [Paul's hands and feet were bound to prevent him from creating fraudulent effects. A singer and drummer continued the ceremony.] . . . Outside a host of other voices seemed to be joining . . . the solitary singer . . . until there was an enormous crash. The *tipi* was shaking as Paul's main spirit helper arrived. A strange grinding noise came from somewhere underground. Blue lights flashed on and off everywhere. . . . I felt something furry walking over my hand. A long tail brushed against me. . . . An icy tentacle of uneasiness wrapped around me. . . . I realized that . . . I was receiving feverish impressions from [a mind that] was not my own. . . . I found myself walking unsteadily down a long, grand corridor. People stood grouped on either side, laughing and chattering. . . . In the small room under the stairs, I saw someone being raped. It was the girl who was sick. . . . Paul stood silently, looking the rapist in the face. . . . Suddenly Paul called to his son to stoke the fire. . . . "I know what is wrong," he shouted . . . . He pointed his finger at the patient's uncle. . . . "The spirits have told me. Confess or you will die." The girl's uncle was trembling. [He acknowledged the rape of his niece and collapsed on the ground] . . . [Paul] laid the girl down on the earth and put his mouth to her abdomen. Making a loud, sucking sound, he pulled something out and threw it into the fire, where it sizzled and burned. "She is well," he proclaimed. . . . The girl miraculously improved. . . . Paul stayed several days afterward to do counseling with the whole family, to prevent this taboo from ever being broken again, and to mend the broken pieces of the lives affected.[1]

This story exemplifies a healing narrative of the form collected by folklorists. When people tell such stories to others, they perpetuate folk healing traditions. Analysis of collections of spiritual healing stories allows insights into the universal features and the physiological bases of spiritual healing practices.

The Native American ceremony apparently involved visual, auditory, and tactile apparitions; psychokinetic and extrasensory perceptions; and sleight-of-hand performance and was followed by counseling sessions. The account is similar to the psychic surgery narrative in the introduction and parallels many ethnographic accounts of Siberian and Native American healing and conjuring ceremonies.[2] Observers have described tents shaking magically and have seen shamans sucking out sickness as described in Mehl-Madrona's account. Within many traditions, the shaman is tied up and placed in a small tent, the tent shakes magically, voices emerge, and listeners gain information that is later verified as true, but that was unavailable to those present. Although such shamanic ceremonies are assumed by outsiders to involve deception, people claim to be cured as a result of them.

Ritual perceptions, such as the anomalous perceptions described by Mehl-Madrona, seem to be associated with hypnotic processes. Hypnosis researchers have generated a growing body of research that supports this argument. People who are more hypnotizable are more likely to produce and experience anomalous phenomena. They are also more likely to benefit from brief psychotherapy.[3] In parallel fashion, when a shaman provides therapeutic suggestions to a group of people, those who are more hypnotizable are more likely to benefit.

The physician Mehl-Madrona notes hypnotic processes within Native American methods: "[The healer] Carolyn worked alone with Wesley, using what psychologists would call hypnosis to prepare him for the ceremony. The Arikara of North Dakota have a phrase for Carolyn's brand of hypnosis; literally it means 'putting him to sleep so he thinks he sleeps when he's really awake.'"[4]

Mehl-Madrona portrays relationships between shamanism, altered states of consciousness, anomalous experience, and hypnosis. In this chapter, I review diverse bodies of evidence that support this linkage. Altered states of consciousness are defined in terms of brain activity. Brain functioning governs anomalous experience. These factors are linked to hypnotic processes that are instrumental in generating wondrous healing. These assertions can be empirically tested, and I review evidence derived from a study of folklore data (like the story provided by Mehl-Madrona) that supports the claim that spiritual healing often involves hypnosis.

Altered States of Consciousness and Brain Activity

Defining altered states of consciousness requires a review of present knowledge regarding normal consciousness. A general assumption is that the brain areas that are most active during a specific activity are related to the performance of that activity. Differences between conscious and unconscious actions can be distinguished by brain-scan patterns. Even the process of quiet inner speech is associated with specific centers of brain activity.[5]

Both animals and humans exhibit brains waves, eye movements, and muscle twitches that clearly differentiate waking, non-REM sleep, and REM sleep. The evidence supports the argument that the modern human brain evolved from that of ancient primates through a series of stages. Within all mammals, states of consciousness are controlled by chemical systems in the brain. The aminergic system allows the waking state with its characteristics of focus, response, attentive awareness, thinking, volition, and analytical reasoning. The cholinergic system eventually leads to dreaming and is associated with lack of volition, thinking, and reasoning. The aminergic system is dominant when we are awake, but the cholinergic system's role increases as the day wears on. When our brains produce less norepinephrine, we become drowsy, we daydream, or we become inattentive. Eventually, during sleep, the cholinergic system brings about rapid eye movements and dreaming, a state characterized by associative, uncritical, imaginative, and fabricative thinking. Aminergic and cholinergic systems continually fluctuate, each becoming dominant at different times.[6]

J. Allan Hobson's model of consciousness is illustrated in chart 1. Consciousness is affected by the level of cognitive *activation* (high vs. low), by the *information source* (internal vs. external), and by the systemic *mode* (aminergic vs. cholinergic). Although the variables affecting consciousness are complex, this model allows us to grasp the major features that influence normal and altered states of consciousness. Analytical waking consciousness exists at point A in chart 1: because amines are dominant, volition is high, information is gained from external sources, and the activation of the cognitive system is high. After a subject is awake for the entire day, amines are weakened, enabling acetylcholine to increase. As the amine system declines, typically in the evening, cognitive activation and external attention decline. The point representing an individual's consciousness sweeps downward and to the front-left, toward the center of the cube (point B). The area around point B is associated with dreamless sleep. As acetylcholine becomes dominant and cognitive activation increases, REM dream sleep begins (point C). This enables the amines to rest and regroup, and eventually to cause waking.

Altered states of consciousness occur during any state of awareness located somewhere other than point A in Hobson's model. Hypnotic processes can be defined as neurophysiological responses that are triggered by the environment and are subject to control through training that results in awareness during deviations from the A-B-C pathway. Hypnotizability is the capacity to generate alterations from the A-B-C pathway. Brain functioning during hypnosis differs from that during REM sleep, and people who are more hypnotizable experience altered states of consciousness in response to suggestion. They experience greater fluctuations or deviations from the typical A-B-C pathway, and the boundaries between their states of conscious are

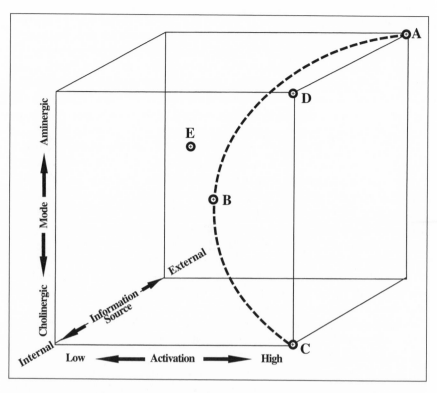

Chart 1: Hobson's Model of Consciousness

less rigid than for the general population. This increased *lability* apparently allows hypnotizable people to affect normally involuntary physiological processes such as blood flow. They also experience unusual perceptions more often than most people and achieve greater benefit from meditation.[7]

Brain Activity and Anomalous Experience

Points that deviate from the A-B-C pathway are of particular interest in the study of religious, mystical, and anomalous experiences. An individual who is experiencing a waking vision is obtaining information from internal sources while highly cognitively activated (point D in chart 1). Meditative visions, daydreaming, and insights associated with relaxation are located toward point E. It is not clear whether alleged paranormal dreams reflect anomalous information sources or whether memory processes generate the perception of an unexplained connection between dream content and reality.

Extrasensory perception, paranormal dreams, visions, and apparitions appear to be related to hypnosis because they seem to involve passage of information between unconscious and conscious areas of the brain. This hypothesis coincides with the research studies finding that hypnotizability is correlated with propensity for religious and anomalous experience.[8] In chapter 5 we shall see that similar forms of anomalous episodes occur among all known human societies, implying a physiological basis. Many anomalous events are like dreaming in that they are perceived internally, yet they differ from dreaming in that they occur less frequently and are more specific to particular individuals.

Researchers have shown that brain lability (variance in electrical discharge), particularly in the temporal lobes, is related to religious, mystical, hypnotic, and anomalous experiences.[9] Arnold Mandell explains the incidence of mystical perceptions as a result of temporal lobe raphe and hippocampal pyramidal cell activity and argues that these episodes inevitably lead to belief in the supernatural.[10] Eugene d'Aquili and Charles Laughlin provide a parallel model based on hemispheric processes: when the left hemisphere is stimulated beyond a certain threshold, the right hemisphere is also activated, and with sufficient driving, simultaneous overstimulation of major brain systems leads to "positive, ineffable affect" synonymous with Freud's oceanic experience, yogic ecstasy and samadhi, Zen satori, Christian *uni mystica,* and other peak experiences directly producing religious belief.[11] Through socialization and practice, the potential shaman gains control over these processes.

Because the temporal lobe region is a later addition within the human evolutionary scenario, John Schumaker argues that elements in this region associated with anticipatory anxiety developed in response to increased intelligence. Temporal lobe activity allows the dissociative capacity to ignore and distort logical interpretations that would be dysfunctional if not curbed.[12] People who report religious experiences are more likely than others to display enhanced temporal lobe signs,[13] and temporal lobe epilepsy is associated with particular types of religiosity, suggesting that this region of the brain affects religious functioning. Kenneth Dewhurst and A. W. Beard surveyed patients with temporal lobe epilepsy; 38 percent showed particular interest in religion after the onset of their illness compared to 8 percent who showed religious interest before the onset of illness.[14] D. Bear and P. Fedio compared scores of twenty-seven temporal lobe epileptics to those of two control groups and found significantly higher rates of religiosity and philosophical interest among the epileptic group.[15] Vilayanur S. Ramachandran and his associates established that one effect of temporal lobe epileptic seizures was to strengthen the brain's involuntary response to religious words.[16] Temporal lobe lability (measured by physiological parameters and by questionnaire) is correlated with frequency of anomalous experience (measured by questionnaire) and hypnotizability (measured by standardized test).[17]

Unusual, seemingly anomalous experiences can be "evoked by either surgical stimulation of the hippocampus and the amygdala or by spontaneous discharges within [the temporal lobe] region."[18] Such experiences include déjà vu, formed visual hallucinations, sense of the presence of a spiritual entity, mystical and paranormal experiences, perception of unusual smells and anomalous voices or sounds, vestibular movements, and anxiety. Normal individuals, as defined by Minnesota Multiphasic Personality Inventory (MMPI) criteria, who display higher lability scale scores are significantly more: "(1) suggestible[19] according to Spiegel's Hypnotic Induction Profile;[20] (2) subject to dissociative states;[21] (3) prone to subjective paranormal experiences and beliefs;[22] and (4) likely to endorse multiple childhood fantasies and imaginings according to the Wilson-Barber scale."[23] People with elevated lability scores also tend to misinterpret odd or infrequent events[24] and to modify their memory of their experiences,[25] characteristics which allow anomalous experiences to be shaped into religious forms.

S. G. Tiller and Michael Persinger argue that the information flow between the brain's hemispheres creates the sense of "presence" that is the basis for awareness of spiritual entities. They have stimulated this sense of presence by firing magnetic fields (1–2 microT) into the temporal lobe area.[26] Persinger's discussion of information flow coincides with Ernest Hartmann's research regarding boundaries in the mind. Thin boundaries, associated with greater temporal lobe lability, are also associated with greater information flow. Hartmann devised and validated a questionnaire about boundaries that can predict the propensity for various cognitive strategies, behaviors, and psychological pathologies.[27] People with thin cognitive boundaries (those who are more open, cognitively flexible, mentally fluid, and creative) are more hypnotizable than other people and report higher incidence of anomalous experience. Cognitive boundaries are linked to physiological parameters: people with thin boundaries are predicted to have more pathways or differing routes within neuronal systems so that thoughts, images, memories, feelings, and so on are better able to flow from one point to another. For thin-boundary people, connections are more like networks rather than like single paths.[28]

The theories of Persinger, Ramachandran, d'Aguili, Laughlin, and Hartmann coincide with Michael Winkelman's observation that synchronization of brain hemispheres is common to all forms of shamanic ASC.

> The common psychophysiology of ASC involves induction of a parasympathetic dominant state characterized by synchronized high voltage slowwave discharges from the limbic system and related brain areas. This discharge pattern results in interhemispheric synchronization and coherence, and limbic-cortex integration, altering human functioning from neurophysiological to cognitive levels in ways which permit the emergence of therapeutic human potentials.[29]

Winkelman reviews the huge body of literature supporting the argument that shamanic therapies have physiological bases associated with psychophysiological dynamics.[30] Except for variations due to sociocultural factors, shamanism and deep hypnosis converge.[31]

Brain Activity and Hypnosis

Hypnotic processes, an important feature within shamanic therapies, have physiological bases. Technologies, such as computerized electroencephalographic (EEG) frequency analysis, EEG topographic brain mapping, positron emission tomography (PET), regional cerebral blood flow (rCBF), single photon emission computed tomography (SPECT), and nuclear magnetic resonance imaging, reveal specific characteristics associated with hypnotizability and hypnotic processes.[32] The evidence indicates that highly hypnotizable individuals can focus more effectively, sustain their attention, and better ignore irrelevant stimuli in the environment than less hypnotizable people.[33] They appear to have a more efficient frontolimbic-sustained attentional and disattentional system.[34] This evidence refutes the argument that hypnosis is merely a matter of role playing unrelated to unique brain processes.

Researchers have established that hypnotizability is shaped by both socialization and genetic inheritance. Childhood absorption in dance, drama, cinema, reading, religion, or dissociation due to abuse stimulate development of hypnotic capacities.[35] To study the genetic basis for hypnotizability, Arlene Morgan tested a sample of 140 twin pairs and their families using the Stanford Hypnotic Susceptibility Scale, Form A. Within the 140-pair sample, the correlations for all monozygotic pairs differed significantly from the correlations for dizygotic pairs (0.52 vs. 0.18; $z = 2.02$, $p < .05$). The overall heritability index of 0.64 falls between those usually reported for personality measures (around 0.40) and ability measures (0.70–0.90).[36] These correlations are comparable to those reported by R. M. Rawlings on the basis of an Australian study[37] and by J. D. Duke in a study of waking suggestibility.[38]

Hypnosis and Anomalous Experiences

The link between hypnotizability and altered states of consciousness allows a hypothesized interpretation of shamanic anomalous experience. When the Native American Mehl-Madrona describes a vision and apparitions of voices, lights, and animals, he reveals his capacity to allow information to flow from unconsciousness to consciousness; this implies permeable boundaries and hypnotizability. We might suspect that the rape victim in his narrative also has thin cognitive boundaries because her trauma manifested psychosomatically and was relieved by ritual performance.

Folkloric accounts from all regions and eras suggest that hypnotic processes facilitate psychic experiences. The history of hypnosis supports this claim. Researchers in the 1700s and 1800s perceived Mesmerism (as "hypnosis" was called before the mid-nineteenth century) to be intrinsically linked to what is today labeled as extrasensory perception and psychokinesis. Eric Dingall collected and reviewed a huge body of studies from the nineteenth century indicating that Mesmeric states facilitated clairvoyance, the discerning of objects or events not present to the senses. Researchers experimenting with Mesmeric clairvoyance included the Marquis de Puysegur, Alexandre Betrand, Baron du Potet, British surgeon John Elliotson (founder of University College Hospital), French psychiatrist Pierre Janet, and Nobel Prize–winner Charles Richet. Although all these individuals witnessed clairvoyance, their experiments were not sufficiently replicable to convince critical observers. Some found the effect transient. For example, John Elliotson, who had been positive he had demonstrated Mesmeric clairvoyance over many years, found in 1865 that he could not reproduce the effect.[40]

Although the modern field of hypnosis has divorced itself from these unusual claims, parapsychologists using controlled experiments continue to find a relationship between hypnosis and extrasensory perception.[41] Meta-analysis of twenty-five experiments from ten different laboratories suggests that hypnosis and similar altered states of consciousness facilitate psychic performance.[42] Although we need not accept this body of evidence as compelling, it supports the argument that hypnotic processes have qualities that bring about anomalous perceptions and beliefs, even among researchers conducting controlled experiments.

Measuring Hypnotizability

Hypnotizability can be regarded as a trait, as is intelligence. Just as intelligence occurs outside of intelligence tests, hypnotic effects occur in a variety of situations, not just following hypnotic inductions. Hypnosis occurs in everyday life—perhaps all the time.

Researchers have devised standardized tests that allow them to investigate the trait of hypnotizability under controlled conditions. To test for hypnotizability, the researcher reads a standardized induction script that typically includes repetitive exhortations to the subject to relax while mentally focusing on the speaker's voice. The induction is followed by a series of suggestions pertaining to different tasks. Subjects' scores on the test are measured by the number of instances of compliance with the suggestions. Tests often include commands regarding nonvolitional muscular movements, analgesia, amnesia, and creation of cognitive images, as well as posthypnotic suggestions.[43] The Harvard Group Scale of Hypnotic Susceptibility,[44] for example, includes an induction suggesting deep relaxation, followed by twelve test suggestions: the subject's head will fall forward; the eyes will close; the extended left arm will become so heavy that it will fall down; the right hand and arm will become heavy and unliftable; the interlocked fingers will become tightly bound and inseparable;

the left arm will become rigid and unbendable; the hands held in front will move together;[45] the head will become unable to shake in order to indicate "no";[46] the subject will hear and acknowledge a hallucinated fly; and the eyelids will become so tightly closed that they cannot open. Two other tests refer to events that will occur after the hypnotic session ends: the subject will touch the left ankle upon hearing a tapping noise (posthypnotic suggestion), and the subject will not remember any of these suggestions upon coming out of hypnosis until told otherwise (amnesia). Subjects who comply with all these suggestions are assigned the score of twelve, while those failing to respond to any receive a zero. Subjects whose responses are in the low to moderate range of hypnotic suggestibility generally are aware that their responses were, to a degree, volitional, but highly hypnotizable individuals often perceive that they were acted upon by outside forces.[47] The nonvolitional element within hypnosis makes hypnotic experience somewhat similar to religious experience, particularly when it occurs "spontaneously," i.e., without clear origin.

Many cultures have devised folk methods for testing shamanic suggestibility. Standardized tests for hypnotizability measure some of the same skills. Shamans "prove" they are in trance by their performance. They demonstrate analgesia (by piercing their body and performing other forms of self-mutilation), out-of-body experience (by visiting spirit worlds or traveling to distant places), extraordinary perception (by gaining information paranormally), special nonvolitional muscular behavior (by acting in the culturally prescribed manner: twitching or shaking, for example), and regulation of normally uncontrolled physiological processes (by controlling bleeding or by exhibiting heat immunity or unusual gymnastic and sleight-of-hand skills). Shamanic systems also involve informal tests of shamanic hypnotizability of clients seeking healing. Tests determine the ability to block pain, to stop a wound from bleeding, to experience specific anomalous perceptions, and to follow a particular behavioral suggestion. A person whose self-inflicted wound did not bleed during a ritual would pass the hemorrhaging test, for example.

Although experimental tests of hypnotizability do not capture the "life or death" quality of shamanic suggestion and "are not equipped to reveal the full meaning of hypnotic responsiveness,"[48] they provide a reliable research tool. As a result, much has been learned through using them. We would assume that within most cultures shamanic suggestibility overlaps Western hypnotizability (as defined by standardized tests).[49] The standardized tests may be regarded as culturally specific; they may reflect the Western notion that the hypnotist controls the person who is hypnotized. Modern researchers realize that hypnotizability constitutes a propensity to respond to suggestion within a particular context and that all hypnosis is, in a way, self-hypnosis.

Hypnotizability possesses characteristics that make it to seem equivalent to experiential forms of religiosity. Its incidence follows a somewhat normal statistical distribution in all measured populations. Test scores are relatively stable over the life span, but nine-to-twelve-year-olds tend to score higher, and scores decline gradually with age.[50] In parallel fashion, propensity for religious conversion peaks around puberty as does the propensity to be the focal person of poltergeist activity (a variable probably related to hypnotizability). Hypnotizability is not highly correlated with personality variables, but is related to the capacity for absorption,[51] fantasy proneness, dissociation, creativity,[52] and propensity for anomalous experience.[53] Hypnotizability is also correlated with scores on the Tellegen Absorption Scale and with dissociative ability.[54] Correspondences between hypnotizability, dissociation, and religiosity are sufficiently close that Schumaker posits that they reflect the same process.[55]

As noted previously, hypnotic response may occur without hypnotic induction. Without benefit of induction, hypnotizable people who are provided suggestions may demonstrate analgesia, amnesia, nonvolitional muscular movements, paralysis, and posthypnotic response. Suggestions that elicit these behaviors without hypnotic induction are termed *waking suggestions* because the subject is not in trance. The Pearson's product correlation (r) between hypnotizability and response to waking suggestions is generally about 0.6.[56] Hypnotic induction increases the likelihood that a hypnotizable person will respond to subsequent hypnotic suggestions.[57] Hypnosis is less a specific set of rituals than an activation of a person's hypnotic potential, however achieved. The role of individual differences in hypnotizability is extremely important in defining the domain of hypnosis, and the outcome of hypnotherapy is frequently potentiated by high hypnotic ability.[58]

Schumaker advocates a "global" perspective rather than a limitation to the vantage point of laboratory research.[59] Hypnosis experiments have a distinctly Western orientation: an operator provides an induction to the subject, followed by suggestions; the implication is that the operator directs the subject's behavior. Modern studies indicate that this model is invalid. Hypnotic response does not require specific induction and is extremely varied. Even when there are no repetitive rituals, waking suggestions can be effective. Western hypnosis can be viewed as a secularized reflection of the more dynamic forms of trance possible in cultures where hypnotic training is part of people's socialization. Hypnosis as it was discovered by Westerners was actually a manifestation of a culturally specific expression of dissociative trance.[60]

Hypnosis and Spiritual Healing

Although experimental evidence indicates that some subjects may anomalously affect living organisms under laboratory conditions,[61] the most common explanation for spiritual healing's efficacy involves hypnotic and placebo effects.[62] This theory coincides with findings from the field of psychoneuroimmunology: psychological variables affect the immune system.

Because hypnosis has a physiological basis, it is predicted that spiritual healing reveals universal features because it often uses hypnosis (see chapter 3). As noted previously, hypnosis works best for particular symptoms, when used with a hypnotic induction, and when provided for a person who is highly hypnotizable. Such people, when hypnotized, often report unusual sensations. Spiritual healing accounts should reveal a high incidence of there elements.

We can evaluate this argument by conducting a content analysis of spiritual healing *memorates*. Folklorists define *memorates* as firsthand accounts of experiences that informants believe to be true. Using content analysis, we can determine the frequency of mention of symptoms, practitioners, methods, and sensations related to hypnotic processes within spiritual healing memorates.

HYPOTHESES

The argument that folk healing success is derived, in part, from hypnotic and placebo processes can be evaluated by testing three hypotheses. First, spiritual healing may be most effective for the symptoms that are most amenable to treatment by hypnotherapy: pain, asthma, warts, headache, burns, bleeding, gastrointestinal disorders, skin disorders, insomnia, allergies, psychosomatic disorders, and minor psychological problems.[63] A collection of accounts of successful spiritual healing should contain frequent mention of these symptoms. Second, in both spiritual healing and hypnotherapy, a specific individual induces an altered state of consciousness through ritual procedures and provides therapeutic suggestions. Folkloric accounts mentioning a specific practitioner, ritual performance, and unusual sensations or immediate reduction of pain imply the incidence of hypnotic processes. Third, because hypnotic and placebo processes are shaped and enhanced by belief, any large collection of folk healing memorates should cite belief as important.

THE DESIGN OF THE STUDY

To test these hypotheses, I had students at Elizabeth City State University collect 1,446 folklore narratives between 1988 and 1996. Many accounts pertained to apparitions, extrasensory perceptions, out-of-body experiences, contacts with the dead, sleep paralysis, and psychokinesis.[64] Analysis of these accounts is provided in chapter 5.

Eighty-five of the narratives described healing as the primary experience. Two independent judges coded the eighty-five healing narratives using written guidelines designed to allow the hypotheses to be tested. Discrepancies in coding were resolved by clarifying directives. In summarized form, the guidelines were as follows:

ANOMALSEN. Does the person being healed (the subject) report a seemingly anomalous sensation or body movement, beyond a reduction in pain, as a result of the healing attempt? Examples include perceptions of heat, perceptions of energy in the body, unexplained emotion, uncontrolled vocalization, unintended body movement, or hearing of an apparitional voice.

LESSPAIN. Does the subject report an instantaneous reduction of pain attributable to the healing activity?

HEALER. Does the informant regard a person who is not a religious authority as able to perform healings?

PREACHER. Does the informant think that a religious authority such as a minister, priest, or preacher is instrumental in the healing?

BELIEF. Does the informant mention belief as a factor influencing either behavior or the treatment outcome?

PRAYER. Does the informant mention one individual's prayer, outside of any other ritual?

GROUPPRAY. Did two or more people, outside of a church, pray for the healing of another? The praying people need not have prayed simultaneously, but their prayers but must have been coordinated in some manner.

CHURCH. Does the informant attribute the healing to ritual activity occurring within a church?

HEALRIT. Does the informant attribute the healing to a nonchurch, nonprayer ritual (a ceremonial action with prescribed behavior)? The HEALRIT category excludes PRAYER, GROUPPRAY, and CHURCH. HEALRIT entails uttering a special set of words or engaging in a series of actions believed, within folk tradition, to bring healing, but not regarded as intrinsically therapeutic within standard medical paradigms.[65]

SPONTAN. Did the healing seemingly occur without preparation on anyone's part?

HERBAL. Does the informant describe herbal or folk remedies involving balms, oils, herbs, or other material substances used for therapeutic purposes?

SYMPTOMS. Does the subject identify symptoms? (Symptoms were coded as described by the informant, then categorized by complaint and body part.)

The simultaneous incidence of three elements, PREACHER or HEALER, HEALRIT or CHURCH, and ANOMALSEN or LESSPAIN suggests that hypnotic processes occurred. This simultaneous incidence supports the second hypothesis (that, in both spiritual healing and hypnotherapy, a special individual induces an altered state of consciousness through ritual procedures while providing therapeutic suggestions) because the preacher or healer has stimulated an assumed hypnotic response using a ritual. Although prayer may sometimes induce hypnosis, the analysis plan cannot ascertain which forms of individual or group prayer constitute hypnotic rituals. Occurrence of the BELIEF code supports the third hypothesis (that because hypnotic and placebo processes are shaped and enhanced by belief, any large collection of folk healing memorates should cite belief as important).

Just as hypnosis and placebo processes occur within biomedicine but do not account for all healing in that domain, hypnotic and placebo mechanisms cannot be assumed to explain all spiritual healing. Alternate processes, including anomalous mechanisms, probably play a role. Folklore research methods cannot evaluate religious or paranormal explanations for spiritual healing.

NARRATIVE ANALYSIS

The most common symptoms described by respondents were burns (14 cases) and cancer (12 cases). The narratives also include mention of fever (7), sores or skin disease (7), stomach pain or disorder (7), common cold (6), ear infection (6), headache (6), pain or disability in legs (6), blindness (5), paralysis (5), broken bones (4), chest pain or heart disease (4), chicken pox (4), eye disease or injury (4), physical injury, excluding broken bones (4), kidney stones (3), urination problems (3), sore throat (3), death (3), demons ("symptom" implying a psychological problem) (2), and warts (2). Eighteen other symptoms were mentioned once or twice. Four symptoms in particular (burns, headache, sores or skin disease, and warts) are often alleviated by hypnotherapy. Other symptoms such as stomach pain or disorder, pain or disability in legs, blindness, paralysis, urination problems, and demons may have psychological bases subject to instantaneous cure through hypnosis. These hypnosis-treatable categories account for 57 of the 135 symptoms (42%) mentioned in the narrative collection. This finding provides support for the first hypothesis (that spiritual healing is most effective for the symptoms that are most amenable to treatment by hypnotherapy).

The collection contains twelve instances (14% of the healing accounts) in which practitioner, ritual, and unusual experience motifs appear in the same account. This pattern supports the second hypothesis.

Comparison of two dominant healing systems allows insight into the means by which practitioners use suggestion to cure (see table 1). Healers were significantly more likely to perform nonchurch rituals, to use herbal remedies, to bring about an immediate reduction of pain, and to treat burns. Preachers were significantly more likely to be associated with church rituals, individual prayer, unusual sensations, and treatment of cancer.

TABLE 1

Comparison of Healer and Preacher Healing Characteristics

Motif (CODE)	Incidence	Healer		Preacher	
Church Ritual (CHURCH)*	20	0	(0%)	8	(44%)
NonChurch Ritual (HEALRIT)*	17	16	(84%)	0	(0%)
Individual Prayer (PRAYER)*	23	3	(16%)	9	(50%)
Group Prayer (GROUPPRAY)	13	0	(0%)	3	(17%)
Herbal Treatment (HERBAL)*	12	4	(21%)	0	(0%)
Unusual Sensation (ANOMALSEN)*	17	2	(11%)	7	(39%)
Reduced Pain (LESSPAIN)*	11	9	(47%)	1	(6%)
Faith in Process (BELIEF)	11	2	(11%)	6	(33%)
Burns (Burn)*	14	14	(74%)	0	(0%)
Cancer (Cancer)*	12	0	(0%)	4	(22%)
Fever (Fever)	7	2	(11%)	0	(0%)
Sores or Skin Disease (Skin)	7	1	(5%)	0	(0%)
Stomach Pain or Disorder (Stomach)	7	0	(0%)	3	(17%)
Practitioner Motif Incidence		19	(100%)	18	(100%)

*Difference-of-proportions tests comparing healer and preacher percentages yields $p < .05$ in two-tailed tests.

Source: McClenon (1997b).

Exemplary narratives portray how hypnosis and placebo processes contribute to healing efficacy within these two systems. Coding has been placed in brackets within the narratives to indicate motifs and symptoms. The second hypothesis specifies that a healer or preacher conducts a ritual that reduces pain or produces unusual sensations. Nine HEALER narratives support this hypothesis by mentioning rituals (HEALRIT) reducing pain (LESSPAIN). Accounts involving burns and warts were particularly illustrative of hypnotic processes. Burn patients often describe immediate anomalous sensations and the alleviation of pain.

> Mrs. Watson talked the fire out of my son ten years ago on January 1. Jeffrey, my son, was burned so bad that he was on the verge of dying [BURN]. Not far

away there was a woman who lived alone and seemed strange and mysterious to others. I didn't think she was weird, just different. I ran down to her
little shack as soon as it happened. . . . Mrs. Watson told me that she knew
something about roots [HEALER]. I took Mrs. Watson to Jeffrey. She began
with a real strange ceremony, then she proceeded to talk the fire out of his
body [HEALRIT]. Mrs. Watson placed her hand upon Jeffrey's chest and began to chant some strange words that I had never heard before in my life. All
of a sudden Jeffrey began to make strange noises as if the fire were releasing
itself from his body [ANOMALSEN]. Finally, Mrs. Watson concluded the
chanting and deep praying. Jeffrey, who was on the verge of dying, got up
slowly but was not fully recovered. A big improvement was shown as he
seemed to feel less pain [LESSPAIN]. This was the strangest thing that I have
ever seen in my lifetime.

Various PREACHER, CHURCH, and PRAYER healings also reveal hypnotic
and placebo characteristics. Seven of eighteen (39%) preacher healings, five
of twenty (25%) church healings, and six of twenty-three (26%) prayer
healings induced anomalous sensations.

The guest pastor [PREACHER] perceived that someone in the congregation was
having tremendous stomach pains [STOMACH], and she wanted the person to come
up to the front of the church [CHURCH]. I did go to the front of the church and she
laid her hands on me and prayed against the situation causing me to be in such discomfort. After she had finished praying for me I felt a warmth go through my stomach [ANOMALSEN], but I still was praying myself for surgery [PRAYER]. After I went
home the next day I scheduled a doctor's appointment in my hometown to perform
the surgery. . . . They wanted to examine me about the blood clots and the bleeding
of my intestines. After several tests, the doctor said, "You don't need any surgery. . . .
Everything looked normal." At that moment I realized that my body was healed supernaturally by Jesus Christ.

Twelve healing narratives described folk remedies (HERBAL). Within
three narratives, herbal treatment was associated with hypnotic patterns.
Healers applied a salve while performing a ritual that induced either an instantaneous reduction in pain or an unusual sensation.

Summary of an interview with Miss Brown:
   Miss Minnie was sort of a doctor in those days [HEALER]. She treated every-
body. Miss Minnie came to Miss Brown's house with some stinking ointment
made of black grease, brown leaves, and fatback meat grease [HERBAL]. "She
put the ointment on my legs and then she got up and started to speak in an
unknown language [LEGS, HEALRIT]. She did this for five days straight. As she
was speaking in her unknown language, she constantly swayed her hand in a
circle." Miss Brown's leg felt better every time she did this [LESSPAIN]. Miss
Brown said right here today, her leg has never given her any problems.

We might assume that some herbal healing efficacy is gained through placebo effects even though only one herbal account specifically mentioned belief.

> Back then they didn't have money to go to the doctor. If a child had a fever they would put a potato on the forehead and they would take the pain away [FEVER, LESSPAIN]. For a earache, they would go in the woods and find a bug and take the bug's head off and then drop the juices from his body into the ear to release the earache [HERBAL]. We didn't have medicine or the money to afford to buy something for that earache [EARINFECT].

Overall, the mention of belief in eleven narratives suggests that placebo effects contribute to healing efficacy.

Individual prayer was mentioned in twenty-three cases and group prayer in thirteen cases:

> The doctor told Mom she had cancer [CANCER]. Since she was a strong believer [BELIEF], she prayed and asked others to pray [GROUPPRAY]. The doctor had said he would give her a hysterectomy. Three days before the operation she awoke from a dream and felt like reading her Bible. Then she began speaking in an unknown tongue [ANOMALSEN]. She refused surgery because she knew she was healed. The following week she visited a gynecologist in Virginia. Although a biopsy had shown a growth, [none was now apparent]. The doctor wanted a copy [of the previous results]. He did not believe it.

As in the story above, some people's belief in spiritual healing is so great that they refuse surgery, thinking themselves cured. A North Carolina surgeon informed me that in the previous year two cancer patients had claimed to have been healed due to faith but that, in both cases, their cancer remained. (These cases cause problems because insurance companies refuse to pay for additional tests to determine if the cancer remains.)

Patients' choice of folk treatment is influenced by their symptoms; this reflects the different functions of healers and preachers. Burn treatments were associated significantly more frequently with healers, nonchurch rituals, herbs, and immediate reductions in pain than were cancer treatments. Cancer treatments were associated significantly more frequently with preachers and group prayer than were burn treatments ($p < .05$; two-tailed difference of proportions tests).

It is incorrect to assume that burns are more amenable to cure by lay healers than by preachers merely because lay healers are more readily available. There are many preachers in northeastern North Carolina. Rather, folk belief specifies that particular individuals have the power to cure burns and that this power can be passed on through family members. It appears that burns are particularly amenable to the one-time hypnotic-type treatment that lay healers provide. Cancer is more amenable to the group processes that preachers mobilize more effectively. Various respondents noted the need to seek lay practitioners for burns and traditional religious practitioners for diseases such as cancer.

Some accounts describe events seemingly beyond the hypnosis or placebo explanation.

> In 1952 my brother and I were playing with some reeds we had cut and were using as spears. We were trying to see who could throw them the highest. He threw one and yelled for me to look up. When I did the spear pierced my right eye [PHYSINJ]. I was taken to an eye specialist in Berea, Kentucky, and the doctor told my mother there was nothing he could do but sew up my eye. My mother told him not to sew it up. I was taken home and my Aunt Barbara prayed for me and put a prayer cloth over my eye [HEALER, PRAYER]. When she was through praying she took the cloth off and I could see at once. I was taken back to the doctor and all he could find was the scar and I have had 20/20 vision ever since.

Four narratives report the rapid healing of children below the age of two. In three other cases, individuals are revived after being thought dead. Folklore research methods cannot resolve disputes regarding paranormal claims, lapses of memory, or informant dishonesty. The existence of similar healing claims within the folklore of many cultures suggests a physiological explanation even though the mechanisms regarding some processes are unknown.

Nine informants reported unexpected, unplanned, and spontaneous healings (SPONTAN). If we assume that the following narrative involves an apparition (the case is ambiguous), then it falls in the SPONTAN category:

> Cotrisha A. was in the hospital room the night before her surgery was scheduled. The doctor had told Cotrisha that it did not look well. Cotrisha had cancer of the uterus [CANCER]. The thought of leaving her young children and her husband frightened her. At this time Cotrisha was only 32 years old. Her family had been in her room to visit with her. Her family left and she was in the room by herself. Suddenly at the door appeared a woman. The woman told Cotrisha that she was not sure if she was supposed to be here or not, and asked her if she could pray with her. She agreed and the woman prayed over her and laid her hands on her stomach [PRAYER]. Cotrisha had her eyes closed, and when she opened her eyes the woman had vanished [SPONTAN]. The next morning Cotrisha went up to surgery, but the doctors could not find one trace of cancer anywhere. The doctor told Cotrisha that her uterus was just as a young woman. Cotrisha believes that this person that came to her was an angel of mercy.

Spontaneous healings tend to be associated with different sensations than those that are stimulated by ritual activities. In order to highlight this difference, unusual sensations were further categorized as unusual percep-

TABLE 2

Type of Unusual Experience versus Spontaneity

	Spontaneous Healing		Nonspontaneous Healing (Alternate Motif Involved)	
Unusual Perception*	5	(83%)	2	(15%)
Bodily Movement/Sensation*	1	(17%)	11	(85%)
	6	(100%)	13	(100%)

*Difference-of-proportions tests yield $p < .05$ in two-tailed tests.
Source: McClenon (1997b).

tions (hearing voices, seeing lights, and so on) or as bodily movements (uncontrolled vocalization or motion) and sensations (feelings of warmth, tingling, vibration). Table 2 indicates that five of the six spontaneous healings were preceded by unusual perceptions (83%) while only two of thirteen nonspontaneous healings (15%) were associated with equivalent perceptions ($p < .05$; two-tailed difference of proportions test). Conversely, nonspontaneous healings produced significantly more bodily movements and sensations, perhaps because people react in a prescribed manner when exposed to ritualized suggestion. Spontaneous events are more likely to involve perceptions that originate with unconscious impulses.

Several cases illustrate the occurrence of bodily sensations in both spontaneous and nonspontaneous healings. Church services are likely to stimulate bodily sensations:

> Interviewer comments: Roger hadn't told anyone that he had broken his back when he was in his preadolescent years. Roger had also severely torn the cartilage in his knees during naval service, and torn ligaments in his ankle were still bothering him enough that he kept them wrapped when he had to do strenuous exercise [LEGS]. The past few years Roger had spent considerable time in the naval hospital with back problems [BACK].
>
> Roger's Narrative: I didn't believe that healing through prayer was possible. We have doctors to do our healing, and they have been through years of schooling in order to learn the art of medicinal treatment. . . . The pastor started calling people to come to the front of the church for healing [PREACHER]. First he called for all of those who were experiencing back problems. Next he called for all persons who were experiencing knee problems. Finally he called for all those who were having problems with their feet to come

forward. Then the pastor said the Lord was showing him that these ailments were going to be cured today. I thought the ailments called forward by the pastor were pretty safe ailments to have in the congregation because of their common occurrence in our society. The pastor then said there was one more person that needed to come forward but was too shy to and that therefore that person would receive their healing where they were. The pastor then commenced his healing prayer and I suddenly went numb from my waist down. I caught myself on the chairs in front of me to keep from falling, but almost instantaneously I felt feeling return. I also felt an intense tingling heat settle first in my foot, then in my knees, and finally in my back [ANOMALSEN]. At the same time I heard a voice in my head speak just two words; those words were "Prove Me." . . . Nobody in that church, including my church date, knew of my maladies. I was surprised and confused. I was also sweating very profusely even though it was very cool in the church. For the rest of the week I commenced to prove what had happened to me. I exercised and lifted heavy items without wrapping up or stretching first. I pushed myself hard on purpose to do things beyond what I knew I was capable of. I didn't want this to be true. At the end of the week, I was convinced that something miraculous had happened to me.

This account is typical in that the subject reports body sensations seemingly in response to the pastor's implied suggestions. The following narrative is unusual in that body sensations associated with healing occurred without suggestion:

I was about 16 years old when I began to notice some disfigurement on the right side of my face. It started off with my thinking it was just an allergic reaction to some unknown food I had eaten. I stopped eating chocolate and drinking sodas. That didn't help though. My face was still getting larger. By this time my mom was getting a little nervous so she called and scheduled me an appointment. I went to the doctor and after about a month of testing and the swelling becoming very noticeable the doctors said it was a tumor on my brain causing the swelling [BRAINTUMOR]. After I was told that, it seemed like everything started going down hill from there. I cried day and night. My hair started falling out and I was going blind. I lost 50 pounds going from about 120 to 70 pounds. Three months had passed then the doctor gave me two weeks to live. One week passed and going on that second week I was in bed and I began to feel hot. By this time my eyesight was gone. I opened my eyes and saw this bright blinding light. The brighter the light got the hotter I felt. I got so hot it felt like my skin was melting. The heat was so intense I passed out [ANOMALSEN]. The next morning I woke up and I didn't feel any pain anywhere. I got up and walked over to the mirror with my eyes closed. When I got to the mirror I slowly opened my eyes. My face was back to normal [SPONTAN]. I couldn't believe it. I ran and got my mom to show her what had happened. She couldn't believe it, she grabbed me and hugged and kissed me with tears flowing down her face. I too was rejoicing; the spirit of God came upon me. I danced all over the house for about two hours, my mother said. After finally calming me down, mother took me to the doctor. When I walked in the office he looked as if he had seen a ghost. He ran some tests on me to see if the tumor was still there, but it wasn't. He couldn't find a thing wrong with me.

Induced hypnotic processes tend to generate body movements and sensations. All over the world, people twitch about, utter unusual sounds, and feel anomalous sensations as a result of ritual cues. Unplanned healings, unassociated with suggestion, are more often related to spontaneously occurring apparitions, extrasensory perceptions, and out-of-body experiences. The anomalous experience itself seemingly provides an impetus for healing. The case described in the introduction illustrates this process. The paralyzed woman heard her husband's voice and rose to her feet in response. After experiencing the aural hallucination, she regained her capacity to walk.

Thus, research links hypnotizability with the propensity for anomalous experience. Physiological processes associated with human consciousness may explain this relationship. People with more open cognitive boundaries and greater temporal lability are more hypnotizable and have greater propensity for anomalous experience. The nature of hypnotizability explains recurring features within folk healing traditions. Folklore research allows analysis of hypotheses regarding hypnotic and placebo effects in spiritual healing narratives. The evidence supports the argument that hypnotic and placebo effects contribute to recurring features within spiritual healing accounts. Forty-two percent of the symptoms described in the collection are often amenable to hypnotic treatment. Fourteen percent of the accounts describe special individuals performing rituals that induce unusual sensations or instantaneous reductions in pain, a pattern suggesting hypnotic processes. Emphasis on belief in eleven accounts also supports the argument that hypnotic and placebo mechanisms play a role in spiritual healing.

5
The Seeds of Religion

About four years ago, I was in hospital because I had a stroke. First I had a numbness on the left side of my body and then I was paralyzed and in a coma for two weeks. You might not believe this but I died and went somewhere . . . not on earth. . . . It was so pleasant and quiet. . . . I was somewhere else when I saw all of my relatives that had been dead for years. I was not scared and I felt at home . . . and Momma [who was deceased] came to talk to me about how wonderful it was and how much she enjoyed her life. . . . We talked and just had a wonderful time together. Then Momma told me about how my family needed me and said that if I left them in their time of need that their world would collapse. She told me that I had a responsibility as a parent and husband to my family. I tried to tell Momma that I wanted to stay with her but she reminded me that I was being selfish. . . . [She wanted me to return.] . . . I did as she said and here I am today. This may not sound believable but it is true.

This near-death account was collected and transcribed by a student in northeastern North Carolina. The student reported that hearing the story increased her belief in life after death. Because people from a wide variety of cultures provide similar accounts, it seems likely that these forms of experience have a physiological basis and that they contribute to similar beliefs about the afterlife all over the world. I argue that the commonly occurring anomalous experiences are "seeds of religion," providing a foundation for belief in spirits, souls, life after death, and magical abilities.

Experiential Source Theory and Cultural Source Theory

The student's argument that her belief was influenced by the story coincides with folklorist David Hufford's experiential source theory: particular experiences are hypothesized to provide a source for folk religious traditions.[1] The theory is associated with three hypotheses: that certain forms of experience have universal features reflecting their physiological basis (similarity); that these episodes occur frequently enough and have sufficient impact that experiencers tell others of their perceptions (frequency); and that stories of these episodes stimulate sufficient interest that they are believed and retold (rhetorical quality), thus creating folk religious traditions.

The ritual healing theory coincides with the experiential source theory because it posits that therapeutic rituals select for the genotypes that allow anomalous experiences that generate religious belief. We can determine the similarity, frequency, and rhetorical quality of specific anomalous experiences by analyzing survey and folkloric data. If certain forms of experience fulfill the conditions established by the three hypotheses, they should be considered a basis for recurring forms of religious belief.

The experiential source theory has been devised in opposition to a prevalent assumption within the social sciences: that anomalous experiences are products only of the experiencers' culture. Hufford refers to this assumption as the cultural source theory. Although the cultural source theory is axiomatic within the social scientific study of religion, it is rarely articulated. (I discuss a permutation of this theory in the concluding chapter.)[2] Cultural source theorists often portray anomalous experiences as products of deprivation or other social pathologies.

Survey evidence refutes cultural source arguments and supports experiential source hypotheses. Religious, mystical, and anomalous experiences are more widespread than many researchers assume.[3] Over one-third of all adults in Britain and the United States claim to have had religious experiences, and about half of all Americans report anomalous episodes.[4] This evidence supports the second of the experiential source hypotheses, the frequency criteria.

The Religious Experience Research Unit in Oxford has collected and examined accounts of religious experience that it has received as the result of surveys and appeals through the mass media.[5] As with other religious-experience and mystical-experience studies, about one third of respondents to their surveys answer affirmatively. Religious, mystical, and anomalous experiences are more likely to be reported by respondents who are female, younger, better educated, and of higher social status.[6] The profile of those most likely to report religious and anomalous experiences does not indicate deprivation or pathology as hypothesized by cultural source theorists.

Filter questions used to elicit responses regarding religious or anomalous experience have an important impact on the type of narrative obtained by a researcher. The filter question of the Religious Experience Research Unit asks "Do you feel that you have ever been aware of or influenced by a presence or power, whether referred to as God or not, which was different from your everyday self?" David Hay analyzed responses to this question.[7] Among those providing an affirmative response, 23 percent believed that there was "a power controlling and guiding me," 22 percent referred to an "awareness of the presence of God," and 19 percent recalled "a presence in nature." Respondents also mentioned "answered prayer," "experience of a unity with nature," "ESP, out-of-body, visions, etc.," "awareness of an evil power," and "conversion." This evidence does not establish whether these forms of experience have universal elements (I will argue later that ESP and out-of-body experiences *are* universal). Any question referring to "God" might bias responses because the notion of God varies from culture to culture.

Andrew Greeley used the "mysticism" question: "Have you ever had the feeling of being close to a powerful spiritual force that seemed to lift you out of yourself?"[8] Over a third of respondents to various national surveys in the United States responded positively to this question. The question has also been used in national European surveys and in my survey of American scientists.[9] Affirmative response is significantly and positively correlated with education, social class, and psychological well-being. Unfortunately, the experiences people describe in response to the filter question vary greatly. Although definitions differ, mystical experiences typically refer to a *union* with a spiritual force or with nature. Religious experiences often refer to a *communication* or *contact* with that force. Greeley's "mysticism" question elicits accounts of mystical, religious, and anomalous episodes. A content analysis of open-ended responses to Greeley's question indicates that 72 percent of respondents who answered the question in the affirmative referred to experiences of a psychic or conventional religious nature while only 5 percent of the reported experiences were judged to be of a mystical nature.[10] A small study, which also asked follow-up questions, found that eight of fourteen people who claimed to have "felt close to a powerful spiritual force" (Greeley's question) had actually not (in the opinion of the researcher) had that experience when their descriptions were closely analyzed.[11]

My collection of memorates suggests that mystical experiences play relatively little role in the generation of folk religion. I used the filter question, "If you have ever had a very unusual experience, would you describe it briefly?" In response, people provided few mystical accounts but many descriptions of anomalous experiences: sightings of apparitions, ESP, out-of-body experiences, sleep paralysis, and synchronistic events. No respondents mentioned coming into contact "with a powerful spiritual force" that seemed to lift them out of themselves.

Yet Greeley's ESP and "contact with the dead" questions allow researchers to estimate the frequency of these common anomalous experiences.[12] Various European and American national samples have been asked about déjà vu, extrasensory perceptions, clairvoyance, contacts with the dead, and the "being close to a powerful spiritual force" experience mentioned

TABLE 3

Percent of Respondents Reporting One or More Anomalous Experiences

	Déjà Vu	_ESP_	_Contact with Dead_
American National Samples:			
United States (1973)	59	58	27
United States (1984)	67	67	42
United States (1988)	67	65	40
United States (1989)	64	58	36
European National Samples:			
Great Britain		36	26
Northern Ireland		24	12
Republic of Ireland		19	16
West Germany		35	26
Holland		27	11
Belgium		18	16
France		34	23
Italy		38	33
Spain		20	16
Malta		28	19
Denmark		14	9
Sweden		23	14
Finland		35	15
Norway		18	9
Iceland		33	41
Average for Western Europe		**32**	**23**

Sources: Greeley (1975, 1987), J. W. Fox (1992), Haraldsson (1985).

previously.[13] Among most surveyed groups, the more educated and higher-status people report more anomalous experiences, and the incidence of these episodes is not correlated with psychopathology.

Table 3 lists the incidence of reports of déjà vu, ESP, and "contact with the dead" experiences by respondents in European and American national samples. Although national groups vary regarding incidence, all report each of the forms of experience at levels sufficient to have impact on folk religious traditions, and folklorists have documented supernatural traditions reflecting these experiences in all cultures. These results could be replicated by surveying any population and asking about anomalous experience. This evidence supports the experiential source theory because the distribution of these experiences does not directly reflect deprivation or cognitive incapacity, because the frequency is sufficiently great to affect folk traditions, and because all surveyed groups report similar forms of experience, implying physiological bases.

The experiential source model portrays a chain of events leading to the creation of a folk religious tradition. A person has an experience, interprets the episode, and describes the memory to others. Some of these stories are repeated, and these oral transmissions constitute the society's oral tradition. Any single account's impact depends on its rhetorical quality, which affects the degree it would be retold. Some types of experience occur frequently but have little impact on religious belief (déjà vu, for example). Conversely, the narrative of a single anomalous event might be widely distributed and have powerful effects on future religious traditions. The return of Jesus, which many scholars consider to have been apparitional, is an extreme example. Folklore collections provide only a vague idea of how often people perceive each experiential form, because respondents to a folklore collection's request provide accounts they feel are worth telling (good stories), only rarely focusing upon the relative frequency of the experience.

Evaluation of Anomalous Experience Narratives

I became aware of the most common forms of anomalous experience when I surveyed a sample of scientists from the American Association for the Advancement of Science (AAAS) in 1981. The 339 respondents were all AAAS Section Committee representatives or council members. The data indicated that these scientists tend to reject paranormal claims, to have less belief in extrasensory perception than other scientists, and to report far fewer anomalous experiences than average citizens. Yet 10 percent of these scientists wrote narratives describing anomalous experiences similar to those of lay people: sightings of apparitions, waking extrasensory perceptions, precognitive dreams, and out-of-body experiences. Among all the variables measured by the questionnaire, "frequency of anomalous experience" was most predictive of "belief in ESP."

While I was a visiting professor at Northwestern Polytechnic University in Xi'an, China, I surveyed a random sample of Chinese university students regarding anomalous experiences. Previously I had discussed my research plans with experts in Taiwan. They had told me that there was little chance that Chinese students would admit to any anomalous experiences because they were all members of the Communist Youth League. Anomalous perceptions did not resonate with the doctrines advocated by the Communist Party. A common social scientific assumption is that social and political climates govern the *incidence* of anomalous experiences as well as the *reporting* of them. The knowledgeable experts assumed, as I did at the time, that the Chinese students would report few anomalous episodes because they had few experiences to report. We thought their lives would not be conducive to having such experiences because of the Cultural Revolution and other historical events. Chinese people in Xi'an who reported no anomalous experiences tended to agree with this assumption. As one student later wrote on his questionnaire, "Life as a student is humdrum and tedious. I don't think we are allowed to have such experiences." To my surprise, a high percentage of Chinese students described apparitions, extrasensory perceptions, out-of-body experiences, psychokinesis, and other anomalous events.

I went on to survey other student groups: dormitory residents at two other Chinese colleges in Xi'an, (1986); dormitory residents at the University of Maryland, College Park (1987); students at Elizabeth City State University, North Carolina (1988); students at Tsukuba University, Japan (1989); and students at the University of North Carolina, Greensboro (1990).[14] All these groups were polled using Greeley's ESP, clairvoyance, and contact-with-the-dead questions as well as questions referring to sleep paralysis and out-of-body experiences. Table 4 provides the percentage from each group claiming one or more of the surveyed types of experience. This evidence supports the experiential source theory: all groups report all types of experiences at levels that would impact folk beliefs.[15]

While teaching at the University of Maryland, I assigned students to interview selected survey respondents to determine the degree to which structured questionnaire responses reflected actual memories. The students found that virtually everyone who claimed an experience in the questionnaire could describe a corresponding event, and many described other incidents that the formal questionnaire did not elicit. They found that Greeley's "contact with the dead" and "contact with a powerful spiritual force" questions stimulated unexpected responses. For example, some people claiming "contact with the dead" experiences merely described vague sensations or impressions, events that would not be considered anomalous by a skeptic. The "contact with a powerful spiritual force" question stimulated response from "born again" Christians who provided no mystical or "out-of-body" element within their narratives. Except such "false positives," all people who reported experiences on the quantitative questionnaire provided anomalous accounts to back up their claims.

TABLE 4

Comparison of Sample Surveys
Percent Reporting One or More Anomalous Experiences

	N	percent responding	Déjà Vu	Sleep Paral- ysis	ESP	Contact with Dead	OBE	Belief in ESP*
Elite American Scientists	339	71	59	**	26	10	20	20
University of Maryland Students	214	42	89	37	44	25	27	66
University of N.C., Greensboro, Students	532	98	86	32	42	20	18	60
Elizabeth City State University, African- American Students	391	99	80	50	35	25	18	68
Chinese Student Random Samples (three colleges)	314	40	64	58	71	40	55	76
Tsukuba University, Japanese Students	132	33	88	50	35	10	13	61

* Percent considering ESP "a fact" or "a likely possibility."

** Elite scientists were not polled about night-paralysis experiences. The out-of-body (OBE) question for elite scientists was the one used by Greeley (1975), which differed from the OBE question used on the other surveys.

Sources: McClenon (1993, 1994: 27).

The questionnaires also collected qualitative accounts of anomalous experience. Each questionnaire included the item "If you have had a very unusual experience, would you describe it briefly?" The question allowed respondents to describe specific anomalous experiences. This procedure allowed me to collect accounts from members of a variety of cultures and subcultures: Caucasian-Americans, African-Americans, Japanese, Chinese, and American scientists. Caucasian-Americans, Chinese, and Japanese groups reflect different cultures, while African-Americans in northeastern North Carolina constitute a subculture.

I found that specific forms of anomalous experience had features that were consistent across cultures. I organized the stories into categories and noted that accounts in each group were so similar that they could have come from any culture: sighting of apparitions, waking ESP, paranormal dreams, out-of-body experiences, sleep paralysis, psychokinesis, and synchronistic events appear to have universal features. Because these experiences occur in all cultures, we should assume that they have physiological bases. I provide example cases from the surveyed groups:[16]

Apparitions   *African-American Student:* One night as I lay restlessly sleeping, I had a visitor. My grandmother, dead for three years, was standing by my bed, stroking my head . . . assuring me that everything was going to be all right.

Caucasian-American Student: Two years ago on the anniversary of [my father's] death . . . I prayed at his grave for a few brief moments and I turned towards my car. I saw him sitting in my car, but the closer I got the less I saw, and when I was close enough the vision disappeared.

Chinese Student: My whole family [lived by a cemetery. One night] the other people were fast asleep. . . . I saw a female standing in the door. I looked at her for a long time . . . [she vanished, but later reappeared]. It wasn't an unreal image. It is a real fact.

Japanese Student: I once saw the ghost of a dog we had owned, sitting on top of a grave at *obon* [the season when dead spirits return to visit the living]. My mother and younger sister also said that they had seen its ghost too. . . . I once saw lots of spirits lined up on the assembly stage in our junior high school.

American Scientist: When I was moving out of my apartment after graduate school, I thought I saw an apparition in the living room as I lay in my bed. I had rented the apartment furnished and the lady who rented it to me recently died. It appeared to me that the apparition might have been of this deceased lady checking to see that I did not take any of her possessions when I left for a new residence. However, I would not swear I saw a ghost as other explanations are possible.

Paranormal Dreams   *African-American Student:* Once I dreamed I was going to a funeral and the very next day my great-grandmother died. The funeral was just like in my dream. . . . I just kind of sat there and thought to myself that I had already been once before.

Caucasian-American Student: I dreamed my grandmother died and knew when, where and how. I was also aware of the last time I was to see her and knew I would not be at the funeral.

Chinese Student: Once I had a terrible nightmare at night [that I would be hit by a car]. I was hit by a running car the next morning.

Japanese Student: I often saw dreams which I took to be precognitive. . . . For example, I saw an erupting volcano (about a year before Mt. Mihara erupted); an airplane crashing (which I later took to be the JAL crash). . . . I sometimes saw test problems in my dreams (which actually turned out to be problems on the real test when I took it.)

American Scientist: On three occasions in my life (early, not recently), I dreamed of events that occurred in advance of their actually happening. Two of the events involved the deaths of individuals—one expected, but not imminent; the other, unexpected.

Waking Extrasensory Perception *African-American Student:* One day I received a phone call at 5 A.M. informing me of my grandfather's death. I knew that the call would inform me of his death prior to the call. I could feel something was wrong.

Caucasian-American Student: Sometimes I can tell when things are wrong back home without coming in direct contact with them. I call home when I have a "feeling," and Mom tells me something bad has happened such as the time my grandfather died, my sister broke up with her serious boyfriend, and Dad was very sick.

Chinese Student: During my study in the university I am far away from my family and have no relatives here. But quite a few times I have had the premonition that something wrong had taken place in my family. I felt very upset and eager to go home. When I was actually at home, I found either my parents had been ill or [that] there had been something wrong with them.

Japanese Student: When I was at home, some thirty seconds before the telephone would ring, I would know that the telephone was going to ring and whom it was from.

American Scientist: Recently, at dinner (7:30 P.M.), I "knew" that an ill friend of friends (whom I knew only casually) had died. Two days later, we learned that he died that evening, when I had the "feeling."

Out-of-Body Experiences *African-American Student:* I was in my bedroom . . . [when] I began to feel my body lift off the bed and begin to float, but my physical body was still on the bed, yet I was up in the air looking at my body on the bed.

Caucasian-American Student: I was mentally "out of it" and saw myself standing with two friends. I felt I was seeing this several feet off the ground—not "in my body."

Chinese Student: Often I thought as if I were not me. I am another person who is looking at "me." I can feel what "she" or "he" feels. I don't know what was the reason of this sense. . . . I don't think it is possible; perhaps it is better to say that I don't wish it is possible.

Japanese Student: Once when I was meditating, I had the fleeting feeling that I was seeing myself from outside my body.

American Scientist: A family member was ill and hospitalized. I "kind of went into a trance" [and] "traveled" in my mind 400 miles to the hospital where I had never been, looked down into the operating room, saw her there at the beginning of the surgery. As the surgeon prepared to make the incision on the right side, I said to him (in my mind) "no—it's on the left side." The surgeon changed over, made the incision. . . . When I received information about the surgery, I asked, "Which side was involved?" I was told, "they finally decided it was on the left side." I understand that this kind of ethereal travel is possible.

Sleep Paralysis   *African-American Student:* Once a demon visited me. I was lying in bed paralyzed and I began to pray the Lord's prayer and then, and only then, did the thing vanish.

Caucasian-American Student: The most unusual experience I have had was waking up one night and being terrified and unable to move. I felt there was someone in the room with me. It seemed to last a long time.

Chinese Student: I dreamed of an old friend who had died several years ago. . . . I was woken up. I felt I couldn't move though my mind was awake. I tried to take up my hand but it was futile.

Japanese Student: I suddenly became aware of an orange-colored light covering the whole side of the room beyond the foot of my bed. . . . I tried to get up but my body wouldn't move. (I think I was barely able to move my fingers.) . . . I tried with all my might to cry out. . . . But even then, I could manage no more than a groan.

American Scientist: (Scientists were not surveyed regarding such episodes and did not provide an example case.)

Synchronistic Events   *African-American Student:* Mrs. Perkins and her husband put their house up for sale to move to the country. The house she would be moving into was a large two-story house. . . . Weekends were devoted to stripping wallpaper and paint. Mrs. Perkins stated that one Saturday afternoon her husband was stripping the layer of wallpaper in the dining room. . . . He pointed to some writing. . . . There was a name and a date of January 23 on the wall and the name written on it was theirs, Perkins, and the date January 23 was her birth date.

Caucasian-American Student: One morning a couple of years ago, a bird flew into (hit the glass of) our bathroom window. I mentioned to my parents that a teacher I had in high school once stated that an old myth says that when a bird hits your window a relative is supposed to die. Later on that day we received a phone call telling us that one of my father's uncles had died that morning.

Chinese Student: Two days before my mother's death, in the morning, a lot of flies flew into the kitchen of our house. There were very many. I felt very curious. . . . Two days later, my mother passed away. I feel this was very strange.

Japanese Student and *American Scientist:* [Although I lack cases representing these groups, synchronistic events seemingly occur in all societies.]

The common features in these accounts suggest that these forms of experience have physiological bases. Alternate explanations, although attractive
to some cultural source theorists, cannot explain why extremely diverse
cultures provide accounts with equivalent forms. This evidence supports
the ritual healing theory: apparitions, paranormal dreams, waking ESP,
sleep paralysis, and synchronistic events generate folk beliefs that provide
the foundation for shamanism.

From 1988 to 1996 my Introduction to Anthropology students at Elizabeth City State University (ECSU) in northeastern North Carolina collected
folklore narratives as part of an ethnographic research project. They were
instructed to interview family members, neighbors, and friends and to
record responses to the question, "If you have had a very unusual experience, would you describe it?" Informants were not restricted by topic but
were urged to provide any narratives they regarded to be of interest, including religious experiences, general folklore, and oral history. Interviewers
were directed to transcribe respondents' exact words, attempting not to
bias responses. The use of local students for gathering sensitive oral accounts is standard within folklore research; it is more valid methodologically than using nonlocal professional interviewers for this task.

The students collected 1,446 narratives; the accounts were coded regarding the experiencer's gender, age, race or ethnicity, and occupational status,
and whether the narrative was firsthand, secondhand, or *folkloric* (more
than secondhand). Anomalous memorate categories were devised to reflect
forms described in the literature.[17] This data allows analysis of the most
prevalent forms of anomalous experience, the types that have the greatest
impact on folk religious belief. Table 5 lists the frequency of the various categories of narratives.

Apparitions

Apparitions consist of perceptions, thought to be exterior to the observer, that have anomalous qualities. Sightings of apparitions were the
most prevalent subject of the accounts, representing 34 percent of the total. A relatively large minority of people in any society will report having
seen an apparition. The Society for Psychical Research found in 1890 that
out of 17,000 British survey respondents 1,684 (9.9%) reported hallucinations.[18] A combined sample collected at the same time in France, Germany, and the United States found that 12 percent of respondents claimed
experiences with apparitions.[19] Later researchers determined that the
propensity to report such episodes had not declined. In 1948, 217 of 1,519
British respondents (14.3%) reported hallucinations.[20] In a 1990 British

TABLE 5

Types of Narrative within Northeastern North Carolina Collection
Prominence within Narrative

Anomalous Experience Memorates:	Primary	Secondary	Total	% of Reports
Apparition	496	147	643	34.4%
Paranormal Dream	157	22	179	9.6
Psychokinesis	96	73	169	9.0
Healing	85	29	114	6.1
Rootlore	87	6	93	5.0
Sleep Paralysis	71	16	87	4.7
Waking ESP	59	22	81	4.3
Synchronistic Event	42	17	59	3.2
Misc. Paranormal	33	11	44	2.4
Occult Event	30	13	43	2.3
Out-of-Body Experience	30	5	35	1.9
Unidentified Flying Object	29	2	31	1.7
Total Anomalous Types:	**1215**	**363**	**1578**	**84.4**
Other Types of Report:				
Folklore	91	38	129	6.9
Normal Dream	75	18	93	5.0
Oral History	65	5	70	3.7
Total Other Types:	**231**	**61**	**292**	**15.6**
Column Totals:	**1446**	**424**	**1870**	**100.0**

Source: McClenon (2000).

study, 123 of 840 (14.6%) replied "Yes" to a question about hallucinatory experiences.[21] Of respondents to national surveys in the United States in 1990, 9 percent reported having "seen or been in the presence of a ghost" and 14 percent said that they had "been in a house you felt was haunted."[22] John Palmer's survey of a random sample of Charlottesville, Virginia, residents indicated 17 percent had experienced an apparition and that 7 percent had lived in a house they felt was haunted. Of people who had seen apparitions, 74 percent had experienced more than one episode.[23]

This frequency of experience seems sufficient to affect religious beliefs. It is natural for people who see a dead person alive to devise theories regarding life after death. Even as early as 1651, Thomas Hobbes speculated that the "opinion of ghosts" was a "natural seed of religion."[24] Although most people who experience an apparition believe that they have seen someone who has died, not everyone does. Thirty-seven percent of the North Carolina apparition accounts did not mention a dead person.

Charles Emmons argues that there are "abnormal features of perception" associated with apparitions. He found the same recurring features within both British and Hong Kong Chinese apparitional collections: images disappearing or fading out, insubstantial images, glowing images, special white and dark clothing, sickly or horrible appearance, partial bodies, abnormal walking, and abnormal sounds.[25] He contends that these features are universal because they arise within such different cultures.

The North Carolina apparitional collection also reveals the "abnormal features of perception" listed by Emmons: 161 cases of disappearance of the image, 35 insubstantial images, 86 glowing images, 38 incidents of apparitions with white or black clothes, 11 apparitions that appeared sickly or deformed, 26 with partial bodies, 13 engaged in abnormal walking or floating, and 37 that produced abnormal sounds.[26]

These data suggest a physiological basis for the perception of apparitions. Apparitions may constitute a form of waking dream in which cholinergic neurons fire while the experiencer is awake. Or perhaps the brain fills in blank areas of the eye's field, generating the apparitional images.[27] Many apparitional episodes have a dreamlike quality: people suspend critical judgment, act impulsively, are sometimes consumed by fear, and perceive events that do not correspond to normal waking experience. Some respondents are uncertain regarding their state of consciousness. This uncertainty suggests that they are in a state between waking and sleeping.

The cholinergic-neurons theory does not explain all the features associated with apparitions. People who have seen apparitions sometimes claim to have received information paranormally. Twenty-four percent of Palmer's Charlottesville, Virginia, respondents who saw apparitions and 17 percent of my North Carolina tellers of apparition stories reported having gained extrasensory information as a result of the experience.

One recurring motif is simultaneity between the sighting of an apparition and the occurrence of a crisis, particularly a death. This feature is probably universal:

> All of a sudden my eyes were open and there was this figure. . . . At the foot of my bed stood Grandma Helen! . . . Grandma Helen walked to the side of the bed and looked at me, with a smile on her face. Then she turned around and

left. The next day my family woke up and sat down to breakfast. I began telling them about the strange occurrence the night before. My entire family sat there in disbelief and shock because all of them had the same exact dream, except Grandma Helen had gone to their rooms instead. After we verbalized our dreams, the phone rang. There was complete silence. I felt as if the world just stopped; it was the hospital. Last night Grandma Helen had suffered a stroke and died in her sleep. We were all flabbergasted![28]

The improbability of the coincidence of a death and a sighting of an apparition cannot be determined precisely. In 1894 members of the British Society for Psychical Research determined that the number of apparition-death coincidences within their collection was 30 out of about 1300 cases (a ratio of 1 to 43).[29] They calculated that the incidence of apparition-death coincidences was 440 times the number that would occur by chance.[30] It is not possible to calculate the actual improbability of these coincidences because we do not know the percentage of people who were aware that a death might occur or how often people's premonitions were inaccurate. The North Carolina collection revealed a ratio similar to that of the British collection: one in thirty-six sightings of apparitions were simultaneous with a death.

Other forms of anomalous experience also revealed high coincidence with death: one in eighteen paranormal dreams, one in thirteen occurrences of waking ESP, one in forty-four instances of psychokinesis, and one in eight synchronistic experiences occurred while someone was dying. These levels of improbability do not statistically prove the paranormal nature of such incidents, but they do explain why people find such experiences interesting, particularly when the subject had no knowledge that the dying person was at risk. The improbability of these events contributes to the rhetorical quality of stories about them. Accounts that seem to prove the existence of life after death can be expected to gain greater distribution than more mundane narratives. As part of my North Carolina study, I had six judges rate the narratives, evaluating whether the stories were "interesting" and whether they were "worth retelling." Stories with death coincidences were deemed more interesting, and narratives that involved emotional factors, such as contact with loved ones, were considered the most worth retelling.

Although death was a frequent theme in apparitional accounts involving coincidence with crisis, many episodes pertained to nondeath events:

Several years ago I was awakened to hear my daughter Angela scream, "Mommy, Mommy!" . . . I got out of bed; looked inside and outside the house. She was not to be found anywhere. I went back to bed but stayed awake. I heard her scream at 2:20 A.M. She was at a party some miles away. Two hours later, she came staggering home all bruised from head to toe. While she was at the party, a man grabbed her, locked her in a bedroom, punched her repeatedly and tore her clothing. . . . The time I heard her scream coincided with the actual happening.

Some accounts that do not coincide with a death imply that there is an afterlife:

[A woman was awakened during the night by her deceased husband's voice saying,] "Martha, Martha, don't worry. I'm all right." . . . I was so shaken up that I couldn't go back to sleep. I guess I finally fell asleep about 4:00 that morning. I was awakened about 9:00 by a knock on the door. It was my sister-in-law. She told me that she had just heard that the graveyard where my husband was buried had been vandalized. I found out that my husband's tombstone had been destroyed and the police estimated that the time of the crime was between 12:00 and 1:00 A.M. I couldn't believe what I had heard. It was like my husband was reassuring me that he was okay at the time the crime was being committed.

As with apparition-death coincidences, these perceptions coincide with an event that is thought to have triggered the anomalous perception. Various respondents stated that their beliefs did not create their perceptions, as supposed by cultural source theorists, but that their experiences generated powerful, unshakable faith, as experiential source theorists would expect.

Some apparitions are associated with unusual movement of objects. Although I have not collected any reports of an apparition physically transporting a real object in view of observers (and I know of no such stories within the psychical research literature), various accounts mention physical effects remaining after an apparition's visit. Respondents describe ghosts leaving behind a hat, jewelry, ribbon, and a mark on a calendar:

I put the clothes in the dryer because they were not all dry yet. When I turned around and headed for the living room, I saw it. It was like a real pale-lookin' shadow just standing there in the middle of the hall to the living room. It was just looking at me, and me at it. Then it sort of pointed at the wall where the calendar was and a bright flash of light sort of shot out of the thing and blinded me. When I looked again it was gone and I was so scared I got the hell out of there as fast as I could. [The experiencer found that a particular date on the calendar had been strangely marked. Later she learned that she was pregnant and felt that she had conceived on the marked date.] To this day, I believe that the spirit was trying to tell me I was going to get pregnant on March 21.

Many accounts describe more than one witness perceiving an apparition simultaneously. Twenty-five percent of the North Carolina accounts and 22 percent of Palmer's apparitional respondents reported collective episodes. Some narrators attempt to refute the skeptic's claim that group experiences involve collective delusion. They stress their previous skepticism, their perceptive ability, and their normal state of consciousness. They argue that the experience itself has the capacity to overcome doubt:

> While I was sitting there looking at television, I heard a voice, but I didn't see anything. I thought I was going crazy but out of the darkness I saw my aunt sitting in the chair where she died. I thought I was seeing things because I don't believe in ghosts. When I realized it was my aunt, I called my mother and told her about Aunt Betty. She didn't believe me until she saw Betty's apparition sitting there. I suppose she came back to be with us because she claimed that she didn't spend enough time with her family like she wanted to before she died.

In some accounts, the apparition travels from one room to another, allowing itself to be seen by independent witnesses who collaborate each other's perceptions. An alternate motif involves a witness describing an apparitional image to another person and being informed that the description accurately matches that of a known deceased person. These collectively constructed incidents often impel belief among skeptical experiencers, a process supporting the experiential source theory.

Although apparitional perceptions are not contingent on belief, it is common for only certain people within a group to experience them. Hornell Hart reports that a simultaneous experience occurs in only a third of the situations in which two people are in a position to see the phenomenon.[31] More typically, one person sees the apparition while others do not. Those who have seen apparitions in the past are more likely to see them in the future (see chapter 6). These characteristics support the argument that special physiological processes allow such perceptions.

Paranormal Dreams

Paranormal dreams involve information gained while dreaming that was previously unknown to the dreamer and is later verified as valid. Paranormal dreams were the second most frequent form of anomalous experience reported by the North Carolina respondents. In some cases, experiencers are unclear whether they were dreaming or not. Leea Virtanen provides evidence that coincides with that of parapsychologists and suggests that states of consciousness between waking and sleeping are highly conducive to anomalous experiences.[32]

Although paranormal dreams do not directly support any specific religious ideology beyond a vague notion of predestination or fate, people often make religious attributions. Of North Carolina respondents providing explanations for their paranormal dreams, 45 percent referred to ESP, 30 percent to God or religion, and 15 percent to a deceased person (10 percent of the explanations were coded as "other").

Paranormal dreams often provide more details than waking ESP or apparitions. In many cases, dream information is photographic in that it corresponds exactly with future events. In other episodes, paranormal dream events are symbolic, and in some cases they have a fantasy quality (images are unreal but still convey a prediction). As with apparitions, death is a common topic:

> One night I had an awful dream. I dreamed that my uncle had been killed. I woke up in such a cold sweat. . . . I didn't tell anyone about the dream because I figured it wouldn't happen. Two weeks later my uncle was killed in a car accident. A lawn mower rolled off the back of a trailer and cut his head off. His son was in the car with him and his father's blood spilled all over him. The very dream I had became true.

This account is typical of precognitive dreams in that the experiencer received complete information—it was *photographic,* giving identity, event, and details—but it did not produce sufficient conviction to cause the dreamer to take action. It is atypical in that paranormal dreams often fail to reveal the identity of a person who will soon die.

Some people describing precognitive dreams report remorse because they did not prevent the negative events they perceived. In some cases people have taken actions that seemingly averted negative events about which they had premonitions. Most dreams provide insufficient information to allow such intervention.

People who claim to have frequent paranormal dreams sometimes note special distinguishing qualities within these episodes that allow them to identify the dreams as paranormal. For example, one woman stated that extremely "real" dreams were generally precognitive. Others note symbolic markers. For example, in the Philippines, a dream of excrement means that the dreamer will soon receive money. People who report frequent paranormal dreams often have other forms of anomalous experiences, and they are often labeled "psychic" by others in their family and community. In hunter and gatherer societies, such people might become shamans.

Waking Extrasensory Perception

Waking extrasensory perceptions entail gaining information anomalously while awake. Like paranormal dreams, waking ESP experiences are anomalous, but they do not directly support core elements of most religions beyond a belief in fate or predestination. According to Hindu folk belief, the action of normal consciousness obscures the extrasensory information.

Thousands of parapsychological experiments have been conducted in an attempt to verify the existence of ESP under controlled conditions. Although no perfectly replicable experiment exists, parapsychologists argue that meta-analysis of various research lines verify the existence of anomalous effects.[33] Skeptics contest this claim.[34]

Like those who have paranormal dreams, people who report frequent waking ESP events often experience other anomalous perceptions. The ability to predict the future or to see distant events is valued in all cultures, and individuals who have this ability are often regarded as potential shamans. Shamanic traditions all over the world attribute extrasensory powers to certain individuals, and it seems likely that this belief originated because many people claim these episodes.

Researchers who have collected cases from a variety of cultures have identified patterns within apparition, paranormal dream, and waking ESP accounts. Parapsychologists typically place paranormal dreams and waking ESP in a generic category, *ESP*. The existence of patterns within ESP narratives supports the argument that there are physiological bases for these experiences. The data come from a number of studies in many societies.

Edmund Gurney, Frederic Myers, and Frank Podmore analyzed 712 cases of telepathy collected by the Society for Psychical Research (SPR) from Victorian British respondents (often upper class) during the 1880s.[35] They included only cases provided by reliable respondents who claimed that the time between the anomalous perception and the corresponding event was less than twenty-four hours. Syro Schouten conducted a content analysis of the SPR cases, selecting 562 narratives that included sufficient data for analysis and were spontaneous (not part of a planned effort on anyone's part).[36] The stories included accounts of ESP and of apparitions providing extrasensory information. Schouten also analyzed the Sannwald Collection, which consists of one thousand ESP narratives gathered in Germany between 1950 and 1959.[37] Louisa Rhine collected over ten thousand anomalous-experience reports from the United States during the 1940s, 1950s, and 1960s.[38] Schouten analyzed a 1,630-case sample of Rhine's collection of ESP stories.[39] Virtanen collected and analyzed 865 "simultaneous informatory experiences" gathered in Finland during the 1970s. She defined simultaneous experiences as episodes in which information was paranormally received about an event at the time the event occurred.[40]

These studies reveal patterns associated with the (alleged) extrasensory transmission of information. The patterns give rise to hypotheses that can be tested by analyzing collections of accounts from any society. Rhine, Schouten, and Virtanen tested several hypotheses: that ESP accounts often involve close family members; that ESP perceptions often involve death,[41] that paranormal dreams tend to refer to future events while waking extrasensory perceptions more often pertain to the present;[42] that compared to paranormal dreams, waking ESP tends to be associated with greater conviction (indicated by the experiencer taking action);[43] that paranormal dreams tend to provide more data (number of details, identification of person or event) than apparitions or waking ESP;[44] and that "seriousness of event is negatively correlated with "completeness of data," so that messages involving death typically include less detail.[45]

An analysis of the North Carolina collection verifies these hypotheses. Judges devised guidelines for a content analysis of the collection and coded the narratives. Statistical evaluation indicated that coding reliability was sufficient to support the study's conclusions. Analysis confirms the hypotheses: almost 57 percent of the extrasensory messages pertained to family members. Death was an important theme within extrasensory messages. Eighty-four of the 207 paranormal message cases (40.5%) pertained to death, a percentage within the range noted in the literature (37.3–66.7%). The form of the experience was significantly related to whether the message pertained to present or future events. As predicted, paranormal dreams were significantly more likely (82.2%) than apparitions (27.4%) or ESP (47.3%) to provide the experiencer with information pertaining to future events (see table 6). Also as predicted, the form of experience was associated significantly with the experiencer's degree of conviction. Of individuals who reported paranormal dreams, only 14.9 percent demonstrated conviction by taking action, compared to 33.3 percent of those who received apparitional messages and 47.2 percent of those who reported waking extrasensory perceptions. The form of experience was also significantly related to the quality of information. Only 21.1 percent of waking ESP episodes provided complete information, compared to 55.5 percent of paranormal dreams and 61.8 percent of apparitions (see table 6). The data indicate that for dreams the quality of information is negatively correlated with severity of event ($r = -0.222$, $p < 0.007$, $n = 144$). Cognitive mechanisms within the dreaming processes seemingly prevent the experiencer from gaining information about death.

Because these patterns were found in data from a variety of societies, the findings imply that there is a physiological basis for ESP. The alternative explanation, that cultural factors produce the structural features, seems unlikely to be accurate considering the wide cultural variation. A critic might argue, for example, that people tend to report ESP episodes associated with relatives because proximity allows them to identify coincidences more frequently. This explanation could be true but assumes that respondents are unable to distinguish ESP from coincidence. No survey evidence supports this argument. Regarding conviction, a critic might argue that people are less likely to take action while in bed but this does not explain the qualitative difference in conviction between dreaming and waking episodes. Waking ESP often evokes powerful compulsions to act, while most paranormal dreams are not recognized as paranormal until later. Because dreaming and waking involve different physiological processes, it is logical to attribute these differences to those processes rather than to cultural factors.

Out-of-Body Experiences

Out-of-body experiences entail the feeling of being exterior to one's body. When they occur during a crisis or severe physical trauma, they are often labeled "near-death experiences." People who have frequent out-of-body ex-

TABLE 6

Relationship between Form of Experience and Time, Conviction, Quality of Information

	Form of Experience			
	Apparition	*Paranormal Dream*	*ESP*	*Total*
Time:				
Telepathic (pertains to present)	61 (72.6%)	26 (17.8%)	29 (52.7%)	116 (40.7%)
Precognitive (pertains to future)	23 (27.4%)	120 (82.2%)	26 (47.3%)	169 (59.3%)
	84 (100%)	146 (100%)	55 (100%)	285 (100%)

Chi sq. = 70.46, df = 2, Prob. < 0.0001

Conviction:				
High (experiencer takes action)	10 (33.3%)	20 (14.9%)	17 (47.2%)	47 (23.5%)
Low (experiencer does not take action)	20 (66.6%)	114 (85.1%)	19 (52.8%)	153 (76.5%)
	30 (100%)	134 (100%)	36 (100%)	200 (100%)

Chi sq. = 18.36, df = 2, Prob. = 0.0001[1]

Quality of Information:				
Complete	21 (61.8%)	81 (55.5%)	8 (21.1%)	110 (50.5%)
Incomplete	13 (38.2%)	65 (44.5%)	30 (78.9%)	108 (49.5%)
	34 (100%)	146 (100%)	38 (100%)	218 (100%)

Chi sq. = 16.36, df = 2, Prob. = 0.0003

[1]Values are uncorrected for low expected frequencies in 2 cells.

Source: McClenon (2000).

periences generally form the opinion that they have some type of soul that can leave the body, allowing them to experience out-of-body perceptions; the exact nature of this entity (or entities) varies from culture to culture. Sometimes the experiencer perceives an event that turns out to have occurred at a distant location, causing the episode to have an extrasensory feature.

Raymond Moody contends that near-death experiences have crossculturally consistent elements: they are ineffable (experiencers are unable to describe their perceptions in words); they often include certain motifs such as a noise and a dark tunnel; and experiencers often hear news of their own death, perceive that they have left their body, encounter a being of light, meet other people, perceive a border between normal life and the afterworld, and return to their body. He also discusses telling others and corroboration.[46] Universal features imply a physiological basis and anomalous correspondences between the vision and reality stimulate wonder. When people tell others about their out-of-body or near-death episode, they generate folk traditions regarding life after death. Yet most experiencers return to their bodies without experiencing all these features. Michael Sabom reports that 28 percent of out-of-body-experience subjects in nonsurgical cases encountered "the being of light."[47] Kenneth Ring describes five stages of near-death experience: a feeling of peace, separation from the body, entering the darkness, seeing the light, and entering the light.[48] Although most people do not perceive all these stages, many describe their experience as "more real" than normal reality. The episode may include emotionally powerful features, such as seeing deceased relatives, as described in the account at the beginning of this chapter.

Carol Zaleski argues that near-death stories are "formed in an inner dialogue between the visionary and his culture," which "develops in the telling and retelling until it finally comes into the hands of an author who shapes it further for dialectic, polemic or literary use."[49] She believes that "the otherworld journey is a work of the narrative imagination. As such, it is shaped not only by the universal laws of symbolic experience, but also by the local and transitory statutes of a given culture."[50] Medieval and modern near-death accounts reveal similarities, yet often differ in emphasis: modern accounts are more upbeat and ecumenical. Most modern narratives are firsthand memorates, less shaped by "telling and retelling." We might assume that the "universal laws of symbolic experience" reflect a physiological basis for these episodes.

Near-death experience accounts from preliterate cultures reveal the same universal features, suggesting that Paleolithic out-of-body and near-death experiences followed the "universal laws" and had features in common with modern episodes. For example, a New Zealand Maori woman who encountered her first white person when she was "a girl just over school age"[51] reports a near-death experience unshaped by Western culture.

> I became seriously ill for the only time in my life. I became so ill that my spirit actually passed out of my body. My family believed I was dead because my breathing stopped. They took me to the marae, laid out my body and began to call people for the tangi. Meanwhile, in my spirit, I had hovered over my head then left the room and traveled northwards, towards the Ngati Whatue, Ngapuhi, Te Rarawe and Te Aupouri until at last I came to Te Rerenga Wairua, the leaping off place of spirits.[52]

Although culturally shaped features exist within near-death episodes,[53] recurring elements probably contributed to convergence in concepts of heaven and hell in medieval Christian and Buddhist thought.[54] Medieval accounts from Europe, China, and Japan are remarkably similar and often include narrative features that do not coincide with the prevailing theologies of their time. People from all eras found these accounts interesting and accepted them as evidence regarding the afterlife. The North Carolina research illustrates the impact of near-death reports on modern people. One student ethnographer who documented a near-death experience states: "It made me think that when we die, our spirits or souls really leave the body, and don't just die with our mortal beings. I had always believed in life after death because of religious reasons, but there is a difference in believing and hearing that it is actually possible."

People who report frequent out-of-body experiences develop particularly robust forms of belief. Some gain voluntary control over their ability, attaining the basic shamanic skill of spirit travel. The survey data indicate that the capacity for out-of-body experience is significantly correlated with the capacity for all other anomalous perceptions.

Psychokinesis

Psychokinesis involves perception of physical action that defies normal explanation. It is the third most frequent form of anomalous experience reported by the North Carolina respondents. Astonishingly little academic interest has been focused on it considering the impact that psychokinetic episodes have had on folklore traditions. Academics often dismiss reports of psychokinesis, assuming that people with flawed cognitive capacity have attributed naturally produced, sporadic, or random noises to spirits. For example, Stewart Guthrie argues that people believe that unexplained experiences are created by spirits because they have an innate tendency to anthropomorphize. "Diverse phenomena—a door slamming, a tapping at the window, a missing object, a light in the forest, and much more—can be explained by postulating a human behind them. Humanlike models thus account for a vast array of things and events. They explain much with little."[55] Guthrie's theory seems logical if the researcher does not take into account the specific forms of experience that people actually report. None of the psychokinesis memorates involve merely "a door slamming" or "a tapping at the window." Most accounts are far more complex, and they do not directly support the anthropomorphic explanation:

> In the week after my grandmother's death I, myself, observed many strange occurrences. Blinds covering windows in the house would fly up without being pulled, a lit candle would go out seemingly on its own, and most curiously, the doorbell would ring when no one was there. I was comforted by the fact that I was not the only one who witnessed the occurrences. At various intervals my mother, father, and brother had the same or similar experiences. Often it

would happen when we were all in the same room together, causing my father to cry out jokingly "All right, Effie [the grandmother], that's enough." . . . The possibility of it being more than a joke became more and more real. We realized that maybe, just maybe, Grandma was trying to say good-bye. . . . Those events still go unexplained. . . . There is no rational explanation.

People tend to attribute psychokinetic events to the deceased. Among the fifty-four psychokinesis informants providing explanations, 79.6 percent cited a deceased person. Many respondents claim that anomalous incidents began with some specific person's death. Such episodes have generated folk traditions in all societies, and unusual perceptions are then attributed to spirits even when the spirits' identity is not determined.

One particular day the chairs on the porch just started rocking as if someone was sitting in them. The wind was not blowing and nothing I could see could explain the reason for the chairs rocking. I also have heard dishes rattling in the kitchen but when I went in the kitchen to look, no one was there and the dishes were all in place. I really didn't believe in ghosts but a lot of strange things have happened. I guess I believe that if a person dies and is not satisfied about something then you can experience problems.

When students transcribe their relatives' accounts, they capture the process by which folk belief is constructed and transmitted. Rather than illustrate how people tend to anthropomorphize, the process of collecting the accounts demonstrates how people in groups tend to generate collective interpretations supporting folk beliefs. For example, a student interviewed his mother, who told him

I don't know if you remember this, but when you and your brother were younger, your little red piano played by itself when Mr. Elliot Barnes died. It was about 2:30 in the morning when I woke up and thought I heard something. I turned around and your little piano played three notes, DING, Ding, ding! Right after it stopped, the phone rang and it was Mrs. Annie calling to tell your daddy to come check on her daddy. Earl always took care of him when he was sick. So he got up, put his clothes on, and went down there. Later on, he called me back to let me know that he had passed. I'll never forget those three notes.

The student interviewer writes her evaluation of the account: "I can relate with this story mainly because I know it's true. The story is told by my mother and I vaguely remember the piano playing. The piano was at the foot of my bed, by the door. Because I was half asleep, I wasn't afraid."

Recurring psychokinetic events often are believed to indicate that a particular location is haunted. These incidents will be discussed more fully in the next chapter. Although memorates do not allow us to determine the degree to which folk beliefs are due to anthropomorphism, the tendency for only particular people within all societies to report these episodes suggests a physiological basis.

Sleep Paralysis

Sleep paralysis occurs when a subject awakens, is unable to move, and perceives unusual phenomena. Most scholars do not consider these experiences to be truly anomalous. The body has physiological mechanisms that prevent movement during dreams that may continue to operate even after waking. The physiological state of being between waking and sleeping may contribute to anomalous experiences. The firing of cholinergic neurons while a person is awake can cause dreamlike perceptions, and the paralysis seems equivalent to that caused by hypnotic suggestion.

Hufford was the first researcher to focus attention on the existence of universal features within sleep paralysis memorates. He found common elements in accounts gathered in Newfoundland and Pennsylvania.[56] I have extended the analysis to narratives gathered in China, Japan, North Carolina, and Maryland.

Many respondents attribute their paralysis to spirits, demons, or witches. Experiences may involve difficulty breathing (which induces fear of death) as well as apparitional features, psychokinesis, and out-of-body sensations. The following story from North Carolina is virtually identical to accounts from Asia, where creatures are part of folk traditions:

> In the middle of the night, I felt as if all of my breath was leaving me. I struggled to inhale, but found that I couldn't. I opened my eyes finally and saw a small creature absorbing all of my breath—drawing it from my nostrils and taking it in for himself. I grabbed it by the neck and began choking it and suddenly it disappeared.

People in northeastern North Carolina often attribute sleep paralysis to the work of witches, which are said to "ride" the sleeping person. Some informants report seeing a deceased person during a dream, then finding themselves paralyzed when they awoke. Alternately, some awoke, saw an apparition, and experienced paralysis. Such episodes contribute to belief in life after death:

> When I finally got to sleep, I felt some cold air. I opened my eyes and I was startled to see someone at the foot of my bed. But I couldn't scream. I just laid there looking at this figure. I couldn't tell who it was, then the figure started getting closer and clearer. Then I realized that it was my uncle. He looked just like I remembered him. We just looked at each other. I'm laying there and I can't move, scream, talk, or anything with tears rolling down the sides of my face. He stood there for what seemed to be about five minutes, and faded away.

Experiencers often believe that sleep paralysis is associated with demonic spirits because their breath is restricted and they feel as though their life is in danger. People who report frequent episodes of sleep paralysis overcome their fear, and some claim to be able to induce sleep paralysis at will.

Synchronistic Events, Miscellaneous Experiences, and UFOs

Synchronistic events are cases in which two seemingly unconnected incidents appear to be related. Omens fall within this category. Corresponding items are often symbolic and specific to a society's folklore (a bird flying into a window coincides with a person's death, for example). Synchronistic episodes suggest a hidden structure to reality that is perhaps related to fate. Some forms of synchronistic accounts involve psychokinesis; a clock may stop at the time of its owner's death, for example.

There are miscellaneous other forms of anomalous events, such as sightings of anomalous ball lightning (the lightning appears to be intelligent) anthropomorphic perceptions (the subject sees faces in the clouds, for example), religious experiences, recollections of a supposed past life, and spontaneous spirit possession. In order for the episode to be classified as a miscellaneous anomalous experience the account must contain some anomalous feature.

> On a recent trip to California, my cousin and her family were admiring the clouds. My cousin decided to take a picture to have for a keepsake to remind her of her trip. The plane flew across a patch of clouds that looked like a bale of cotton. My cousin couldn't resist such a wonderful photographic opportunity, so she took a picture of the clouds. Later, in the week she received her developed pictures. The picture of the clouds nearly knocked her unconscious. Among the clouds was an outlined picture of Jesus. He was dressed in a robe. She had the picture enlarged and shows it off in her house. She said the picture was God's way of saying He is watching over us.

Although Guthrie argues that anthropomorphism is the foundation for all religious sentiment, respondents provide only a few accounts that clearly support this explanation. It would seem that other processes are more important. Respondents claiming to have seen unidentified flying objects (UFOs) sometimes report that multiple witnesses also saw the object, that the object made anomalous movements, that it left behind physical evidence, and that the sighting had long-term psychological effects. Although many accounts of UFOs generate belief that the objects are from other planets, these episodes typically do not stimulate religious sentiment.

In sum, surveys of national samples support the experiential source hypotheses. Specific forms of anomalous events occur frequently enough that they stimulate belief in folk religious traditions. All surveyed societies report certain common forms of experience. Random surveys of student populations in China and Japan and in North Carolina and Maryland in the United States provide a collection of accounts that allow comparison. Apparitions, paranormal dreams, waking ESP, out-of-body and near-death experience, psychokinesis, sleep paralysis, and synchronistic events involve common features suggesting physiological bases. These commonly occurring anomalous experiences have rhetorical features that make them likely to be retold. They often pertain to death, include emotionally powerful elements, and involve issues people consider important. Analyses of these data support the argument that recurring anomalous episodes provide a foundation for folk belief in spirits, souls, life after death, and magical abilities.

6

When the Seeds of Religion Sprout

> By the time I was three years old it became very apparent to my parents that they did indeed have a child with a frightening and undesirable trait that caused her to claim the ability to see and hear those who had died. . . . To further complicate matters I would describe people whom my parents had known in the old country and who, unknown to my parents, had died. These phenomena greatly disturbed my parents. . . . I was the one child of the eleven who saw things that no one else could see, who made prophecies that were laughed at but that came to pass, much to their consternation and alarm.[1]

The spiritual healer Olga Worrall portrays the stigma associated with psychic experience. She also describes events that led those around her to believe in psychic phenomena. When she was twenty-one, she and her family began hearing loud knocking from various locations in her room. Soon afterward, she was informed by a psychic that she was a "sensitive." After she got married, she and her husband experienced poltergeists in their first apartment. They moved to a new apartment, and after encountering further unusual experiences, they came to believe they could heal people psychically. Because of their reputation, many people sought their help. Parapsychologists conducted controlled experiments testing her paranormal abilities and reported that these abilities were authentic.[2]

Olga Worrall's autobiography (co-authored with her husband, Ambrose) illustrates how people who experience frequent anomalous experiences affect those around them and are affected by them in turn.[3] Anomalous experiences are culturally interpreted, and stories of them generate folk religious traditions. Surveys, folklore data, participant observation, and anthropological evidence portray how anomalous experiences, the seeds of religion, can sprout and eventually grow into folk religious traditions. The data allow a scenario regarding the Paleolithic origin of religion. Paleolithic people experiencing frequent anomalous experiences gained robust belief in spirits, souls, life after death, and magical abilities. Their beliefs were shaped in conjunction with those around them who also shared in some of their experiences. This led to collective ritual activities which reinforced beliefs and aided in developing more complex religious ideologies.

The scenario explains why some Paleolithic people became shamans. Frequent experiencers tend to have thin cognitive boundaries, and they sometimes report special problems as a result (experiences with poltergeists and hauntings, possession, spiritual crises, and culturally specific psychosomatic infirmities). If they gain control over their anomalous capacities, they may be cured, and they may gain the capacity to heal others. Their socialization process, which involves anomalous experiences, provides them with unshakable faith that allows them to perform rituals that produce hypnotic and placebo effects in others. This process probably occurred in Paleolithic times just as it occurs today. People with thinner cognitive boundaries probably have always been more hypnotizable and have experienced more frequent anomalous experiences, suffered more often from culturally specific disorders, and received greater benefits from shamanic healing than most people do. We can study the processes by which shamanism originated and is continually being re-created using surveys, folklore research, participant observation, and anthropological studies.

Surveys

Surveys of samples from a wide variety of societies reveal varying percentages of people who report frequent anomalous experiences and also portray characteristics of these people. Discussing his random sample survey of residents of Charlottesville, Virginia, and of students attending the University of Virginia, Charlottesville, John Palmer notes

> There is considerable evidence from the survey that most of the respondents who claimed to have had psychic or psi-related experiences apparently had a large number of them. . . . Secondly, there was a tendency for respondents who reported one type of experience to report other types as well. . . . Among the 55 cross-tabulations contrasting the 11 items discussed. . . . 50 of them were statistically significant (i.e. $p < .05$), most highly so.[4]

Other researchers have found parallel patterns. Andrew Greeley explains that a sizable majority of U.S. respondents reported at least one paranormal experience and that almost one-fifth reported frequent paranormal experiences; people reporting one form of experience tended to report other forms.[5] My surveys support Greeley's findings: the incidence of all the forms of anomalous experience are significantly correlated with each other in almost all instances. Frequent experiences were reported by 22 percent of respondents at the University of Maryland; 13 percent at the University of North Carolina, Greensboro; 13 percent at Elizabeth City State University in North Carolina; 54 percent at three colleges in the People's Republic of China; and 16 percent at Tsukuba University in Japan. Eleven percent of elite American scientists from the Association for the Advancement of Science reported frequent experiences.[6]

Researchers have found correlations between certain psychological variables and the incidence of anomalous experience. Sheryl Wilson and Theodore Barber found that 92 percent of their sample of highly hypnotizable women, most of whom described paranormal episodes, considered themselves to possess psychic abilities. In contrast, only 16 percent of less hypnotizable subjects reported such beliefs and experiences. More than two-thirds of these fantasy-prone subjects, and no members of the comparison group, believed that they had the ability to heal someone spiritually.[7] Robert Nadon and John F. Kihlstrom[8] and Michael Dixon and Jean-Roch Laurence[9] replicated Wilson and Barber's study: respondents were evaluated using a "Paranormal Experiences Questionnaire," which measured belief and experiences labeled as paranormal, psychic, psi, or anomalous. Belief in and experience of these episodes are significantly correlated with hypnotizability.

After many replications by other researchers, the link between hypnotizability, propensity for anomalous experience, and belief in the paranormal appears to be well established.[10] Studies indicate a constellation of psychological variables that imply the existence of a *shamanic syndrome* characterized by hypnotizability, dissociative ability, propensity for anomalous experience, fantasy proneness, temporal-lobe signs (measured by questionnaire items regarding unusual experiences associated with temporal lobe epilepsy), temporal lobe lability (measured by EEG), and thinness of cognitive boundaries (measured by Ernest Hartmann's boundary questionnaire).[11] A principal-components analysis of variables such as belief in paranormal phenomena, magical ideation, manic and depressive experiences, and scores on a creative personality scale found that a single factor accounted for 52.5 percent of the variance in one study and 54.2 percent in a replication. Michael Thalbourne and Peter Delin labeled the factor "transliminality," the degree to which there is a gap in the barrier or gating mechanism between the unconscious (subliminal) and conscious mind.[12] Transliminality has been found to be highly correlated with a measure of mysticism, and people who are high in transliminality are more susceptible to incursions of ideational and affective input from subliminal regions.[13]

Hartmann provides a framework that explains these patterns.[14] He hypothesizes relationships between his measure of thinness of cognitive boundaries and a variety of behavioral and attitudinal factors (the cognitive boundaries measure seems similar to transliminality). Cognitive boundaries are genetically based and culturally socialized barriers to the spontaneous

flow of images and information within the brain. People with thin boundaries have the sense of merging with their perceptions. They demonstrate
characteristics related to emotional openness and have greater fluidity of
thoughts and feelings because there are fewer barriers or walls separating
them cognitively from the world. Thin cognitive boundaries allow hypnotic suggestions to affect unconscious processes, a characteristic associated
with certain cognitive pathologies. Thinness also facilitates the flow of
anomalous perceptions from unconsciousness into consciousness.

The shamanic syndrome, which consists of features such as high hypnotizability, thin cognitive boundaries, and propensity for anomalous experience, has costs as well as benefits. Powerless people, in particular, suffer the costs associated with the shamanic syndrome. People who have a
low social position are less able to defend themselves when attacked psychologically. A person required to listen to a superior's frequent verbal
abuse, for example, might develop psychosomatic, even physically based,
symptoms as a result. Because of this mind/body relationship, lower status
people have lower life expectancy and higher death rates than upper status
people. Lower-class people also suffer from higher incidence of mental disorder. Although many factors contribute, the greater incidence of physical
and psychological illness is mainly due to the greater stress associated with
lower status.[15] Powerless people with thin cognitive boundaries are particularly at risk because psychological stress places an added burden on their
immune systems. Research indicates that even among animals social position affects health.[16] Because genetic factors in animals influence animal
temperament, sociability, and independence, coping mechanisms regarding stress in both humans and animals undoubtedly have genetic bases.
Therapeutic rituals could provide particular benefits for less powerful humans with thin cognitive boundaries. In summary, the shamanic syndrome causes people with thin cognitive boundaries to suffer from psychosomatic problems, to benefit from ritual healing, and to become shamans.
Because the research literature does not indicate an inverse relationship
between social class and propensity to experience anomalous events, we
might speculate that, in general, thinness of cognitive boundaries provides
survival advantages—allowing people to devise creative strategies that lead
to their empowerment.

Folklore Research

Qualitative analysis of folklore accounts allows insights into the group
processes surrounding the shamanic syndrome. Anomalous experiences
within small groups reveal a bridge between spontaneous individual experiences and resulting beliefs, and the complex practices and beliefs associated with modern religions. Recurring patterns within folklore data imply
that frequent experiencers often come to believe that they can control
anomalous incidents—that they have magical abilities; that frequent

experiencers facilitate spontaneous anomalous perceptions among those around them, generating belief; and that group occult practices (often led by frequent experiencers) also facilitate sought-after anomalous experiences and generate belief.

Apparitions and psychokinesis are often witnessed by more than one person simultaneously. Such collective incidents increase belief in the authenticity of the episodes. Within the North Carolina narrative collection, 12 percent of apparition experiences involved simultaneous sighting by two people, and an additional 11.2 percent involved sighting by more than two people. Psychokinesis is almost invariably witnessed by all present.

Hauntings are situations in which multiple, apparently related apparitional or psychokinetic experiences occur in the same location. They are often experienced by people within their homes. The word *haunt* comes from the prehistoric German and is derived from a word that was the source of the English *home*.[17] *Poltergeist* is an Old German word meaning "noisy spirit." Hauntings often lead to belief that a spirit is making a particular place its home. Although psychical researchers argue that poltergeist cases tend to have characteristics that are different from those of hauntings (poltergeists more often involve psychokinesis, occur around a specific individual, and include fewer apparitional events),[18] folklore accounts fail to make this distinction. For convenience, I use the term *haunting* for both recurring apparitional episodes and for situations that might technically be classified as poltergeist cases.

Like the anomalous experiences from which they are constructed, hauntings reveal recurring, cross-culturally consistent patterns. Alan Gauld and A. D. Cornell analyzed five hundred haunting and poltergeist cases that vary greatly in temporal setting (A.D. 530 to modern times) and geographical location (Europe, North and South America, Africa, and Asia).[19] All over the world, people report similar forms of recurrent spontaneous apparitions and psychokinesis and attribute these events to deceased people and other spiritual forces. Within my North Carolina primary-experience collection, about 35 percent of the apparitions and 56 percent of the psychokinesis episodes pertained to hauntings.

> When we first moved in, you could hear noises and voices at night, and there were sounds of furniture being dragged across the floor . . . and you could hear water running but when you went to the bathroom door it would stop. Sometimes in the middle of the night you could hear a man's voice calling out. They told us that someone had died there . . . that the man who lived there before us was murdered. My parents went to go see a minister about all the strange happenings and he came out to the house and did an exorcism. . . . In other words, he blessed the house. After that, we haven't had any more problems with noises or voices.

Haunting accounts support belief in spirits, souls, life after death, and sometimes the ability to influence spirits. Many respondents claimed that they were originally skeptical but became believers because of their experiences. I will discuss hauntings further in the "participant observation" section that follows. Because such a large percentage of apparition and psychokinesis accounts involve hauntings and because so many people in all societies have had these experiences, hauntings have probably had significant impact on folk religious traditions throughout the ages.

One form of folklore account, which involves what I label "occult practices," refers to magical actions producing anomalous perceptions outside of organized religion. Many of these stories illustrate an intermediate stage between spontaneous experience and shamanic performance. Occult practices do not require a special practitioner or complex religious ideology, but groups often benefit from the advice or participation of experts. My North Carolina collection contains 136 occult practice narratives including 93 stories describing rootlore, which is a special occult tradition with African origins.[20]

Occult-practice and rootlore stories often describe symbolic or magical actions that generate what we might assume to be placebo or nacebo effects (nacebos produce harm rather than benefit). These narratives almost invariably involve belief in magic and often portray how a nonbeliever comes to accept occult beliefs. Many stories include complex symbolic actions based on a "recipe," or standard formula:

> I was a happily married woman until one day I found out that my husband was cheating on me. . . . I wanted my marriage to last, and I wanted the other women to leave my husband alone. I didn't know what to do. One day I was talking to my grandmother and . . . she gave me a spell to put on him. I didn't believe in "roots," but I gave it a try. I had no other choice. I was desperate. So I did what she told me to do. She told me that when the moon became full on the second Wednesday in the month, I should take two strings of my hair and put them in a glass of water that I had previously drank out of, and make sure that he drinks all of the water. If he did this and if the moon was right, he would never stray again. He drank it, and to this day my husband has never been with another women.

These accounts differ from spontaneous anomalous experience narratives in that they often refer to folk traditions advocating specific rituals. Expert practitioners sometimes gain reputations:

> Your Granddaddy Ervin was a warlock. He messed around in voodoo so heavily that many people was afraid to get him upset. If he couldn't have things his way, then he would conjure up a spell to make things go his way. One day he was arrested for doing someone's taxes wrong. He went to court and he did some kind of spell and the judge threw the case out of court. He should have been in prison for life but he managed to escape prosecution.

Various stories describe groups testing magical recipes to determine whether they work. A common form of group occult practice involves subconscious muscular movements and leads participants to perceive that spiritual forces create the effect.[21] For example, people play with a Ouija board, touching a marker they perceive to be controlled by a spirit. A common theme within Ouija board accounts is that participants gain information paranormally, and some stories describe apparitions and psychokinetic effects. Many narratives portray the spirits summoned by Ouija boards as demonic. Storytellers sometimes link Ouija board phenomena with other anomalous claims and argue that their belief comes from experience:

> [A skeptical student was participating in a Ouija board group]. I asked boldly, "What is the name of the girl that I am messing with back at home?" I knew that no one knew the answer because I had not told anyone about her. The "little Ouija Board thing" started moving. It moved slowly to each letter. To my fright it gave me the entire name of the girl including her middle initial. After seeing that, I jumped up and ran out of the room. I was shocked, scared, surprised, and excited, all at the same time. Later that evening, about 2:30 A.M., my phone began to ring. I didn't want to answer it because I had this funny feeling that it was the same girl that was spelled out by the Ouija board. Reluctantly, I picked up the phone and it was her. She said that she had to call because she felt that I had been talking about her.

An alternate method of harnessing subconscious muscular movements involves a group placing their hands on a table and finding that the table moves about as if controlled by a spirit. Asian Taoists grasp the legs of a small chair and believe that the chair's movements spell out messages that come from a deity.[22]

People who report frequent experiences often relate a form of shamanic biography with cross-culturally equivalent features.[23] Like Olga Worrall, they typically report that their first anomalous experience occurred at a young age, describe unusual social reactions, and sometimes say that they experienced medical or spiritual problems (which may have been psychosomatic) and that they eventually overcame their problems by adopting the role of a healer.

Anne, a thirty-year-old, African-American North Carolina informant, described most of these features.[24] Her first unusual experience occurred when she was about five years old:

> I remember my uncle being there and he took me down the stairs so I wouldn't hurt myself. He put me in the living room on the couch, and he put something over me to keep me warm. And in the morning when I woke up my mama said, "What are you doing down here?"
> I said, "Uncle Herbert brought me down here."
> She said, "You don't know what you are talking about." That's all she said,

but she knew that I was sleep walking. They didn't tell me until later, when I was about eight or ten, that my uncle had died when I was two years old.

As she grew older, Anne realized that she often perceived things that others did not. She saw her deceased grandfather and frequently heard the voice of a deceased neighbor. When she was young, her family dismissed her stories as fabrications (Her mother would merely state, "You're always seeing things.") But over time, Anne developed a reputation for having dreams and premonitions that came true. Her many experiences include a precognitive dream of a shooting in her neighborhood, a waking ESP sensation that foretold of a relative's death, a precognitive dream of a wake that corresponded exactly with later events, a precognitive dream of a miscarriage experienced by a woman at her work, a precognitive dream corresponding with her sister-in-law's death in a fire, a precognitive dream regarding a trip to Richmond, a dream of a bald stranger whom she later recognized as her brother's attorney (even though he wore a hairpiece), and a series of precognitive dreams regarding winning numbers in the Maryland lottery (she won small amounts on three occasions).

Anne's near-death vision during a surgery shaped her attitudes toward death:

> In the hospital and while I was on the operating table I think I died. . . . I had an out-of-body experience and it lasted for a couple of hours. . . . Some figures came. . . . They took my spirit up to the upper part of the ceiling, and I watched the doctors working on my body. . . . It was real pretty and quiet. . . . And there was water that was blowing so smooth you couldn't tell that it was moving. It seemed like it was wider than the earth, . . . and there was a figure across the water. . . . I was watching what was going on, and then I heard the doctor say, "OK, we're closed now," and they were saying things that they were going to do. I saw that I had a tube down my throat, and when they were finishing up, this figure from way across the water reached his hand out all the way across the water and touched my hand, and said, "You can go back now, and everything will be alright." When he said that, the figures took me back down, and the spirits were around my body again. . . . I know that it did happen, and it wasn't a dream. I felt the man was Jesus. . . . Since that time a number of amazing things have happened in my life. I've always had the assurance that Christ would take care of me. . . . Now, no matter what happens, I really know he's there for me. It's made a difference for me. . . . I ask him for guidance, and I try to take the steps as he guides me. I don't worry about anything because I don't need to. I know that I'm only going to be here for a little while, and when I go I know there is going to be complete peace. I'm not afraid of death or dying.

Given the frequency and nature of her experiences, it is not surprising that Anne has developed strong beliefs regarding the supernatural. She feels, as a result of specific experiences, that her wishes can have direct effects. This power sometimes has negative consequences. Once she became extremely angry at her supervisor and wished him harm. On his way home he had a motorcycle accident. She also believes, based on her experiences, that her prayers can benefit others. People often ask her to pray for them because they believe, as she does, that her prayers are more frequently answered than those of most people. She believes that she can avert the events in negative visionary prophesies through prayer, and she has prayed particularly hard after being warned through her dreams of potential accidents.

Anne also describes gaining information from symbolic dreams that lead her to contact strangers who need her spiritual assistance. The complexity of these events makes their explanation through coincidence seem implausible. She reads special passages from the Bible for people plagued by visits from unwanted spirits. Her stories are parallel to those of other amateur shamans I have interviewed. They describe intuitive sensations and precognitive dreams that cause them to contact people they do not know in order to provide spiritual assistance. A minister-healer impulsively took an unplanned turn toward a church member's house and discovered a medical and spiritual emergency. A woman described knocking on strangers' doors when she was four years old. She would provide them with information about recently deceased people that was given to her by visionary angels. Some people would react with amazement and others would be alarmed. These cases illustrate some of the ways people who experience frequent anomalous events stimulate belief in others.

Some religious environments foster anomalous perceptions. As with occult-practice cases, those with a propensity for unusual experiences play important roles. For example, a North Carolina respondent describes her perceptions during a church service:

> It was towards the end of the service when the Holy Ghost came in and everyone was shouting and dancing. . . . They were overtaken by the Holy Ghost and unable to stop praising God. A woman in the choir was among them. I looked at her and was saying "Praise Him, Praise Him." Then she fell to the floor, still shouting in the spirit. I went down to keep her head from hitting the floor. She was still shouting and praising God, and I began to pray for her. As I was praying I saw something black come flying out of her mouth and head for me. It didn't scare me. I said, "Jesus," and began praying again for these demons to come out of her. They came towards me but bypassed me and went to the ceiling and began circling to find somebody. But the spirit of the Lord was so strong that they couldn't find a host body. They went out of the roof. . . . When the other choir members went to eat, I stayed with her and continued to pray for and with her. The next day I told her what I saw. She said, "No you didn't, you're just saying that," and began laughing. I said, "I don't lie about the Lord." She said that I was scaring her, and I explained that the Lord drove all those bad things out of her and that she couldn't go back to

doing those things again. I even told her two things that she had done. She asked me how did I know. I didn't *know*—I just *knew*. Then she believed me.

Such stories portray how some people who report frequent anomalous episodes end up fulfilling shamanic roles. They experience apparitions, waking ESP, paranormal dreams, psychokinesis, and contacts with the dead, and they often induce others' belief as a result. Some spontaneously adapt therapeutic, shamanic roles; others are extensively socialized by way of training under expert guidance. People who live close to frequent experiencers may urge them to adopt shamanic roles. Paleolithic societies would have contained such people, and it is easy to visualize how they would have instigated shamanic healing systems because their experiences compelled them to believe in spirits, souls, life after death, and magical abilities. This scenario coincides with what we know about the recruiting, sickness, and initiation associated with shamanism.[25] When the seeds of religion sprout, shamanism results.

Participant Observation

Participant observation of groups that report recurring anomalous experiences allows insights into the manner in which these episodes have shaped folk religious traditions. Sociologists can test hypotheses through field observation to produce qualitative *grounded theories*.[26] The observer engages in a continuous process of theory building and hypothesis testing based on field observation. The grounded theory that results must explain a phenomenon or a concept, must focus on relationships, must be verifiable, and must relate concepts to observations. Grounded theories can be distinguished from other types of theory in that they are related directly to people's real-life experiences and to their shared interpretations of the meaning of those experiences.

HAUNTINGS

Although sociologists often cannot determine the authenticity of anomalous claims, they can note attitudes and behaviors of groups claiming to experience such episodes. I devised grounded theories on the basis of twenty firsthand investigations of hauntings and twenty-five telephone interviews with people claiming to have experienced hauntings. Most of these cases came to my attention when haunted people contacted me after they were referred by a parapsychological research center. These cases are not as unusual as might be assumed. Fourteen percent of the general U.S. population claim to have experienced haunting incidents.[27] I expect that most of their experiences are less problematic than those described by the people in my participant observation and telephone interview samples.

In 1979, the Psychical Research Foundation suggested that I contact a family in Baltimore who had requested help.[28] The family reported hearing whistling and voices, experiencing feelings of being bitten at the legs, seeing beds move up and down, seeing an orange-yellow ball of light, and experiencing other anomalous sensations.

I was not certain how I should interpret their stories. I had never talked to anyone who claimed a haunting experience, and I did not think that a normal person would make such a claim (during the 1970s, the media did not portray hauntings as common or real). I suspected that haunting cases reflected some form of psychopathology. This was the hypothesis with which I began my investigation.

On September 30, 1979, I interviewed the family. They described many paranormal incidents: seeing balls of light and snake-like lights; hearing knocks and unexplained sounds such as voices, whistling, music, drums, and cymbals; feeling numbness, tingling, itching, and burning on the skin; and experiencing a paralyzed, possession-type of feeling. They reported unusual nightmares and strange bed shaking.

The owner, Mr. M. (age sixty-two), and his wife (sixty-three) lived in the house with two unmarried sons, Mick (thirty-six) and Robert (twenty-seven); two divorced daughters, Manny (thirty-three) and Nancy (thirty); and one of the daughter's two sons (nine and seven). Mrs. M.'s retired sister (seventy-two) also lived with the family.[29] After they moved into the house in April 1979, they heard unusual knocking sounds. Mick put out poison because he suspected their house had mice. The poison had no effect.

About a month later, Manny woke from a strange nightmare. She felt paralyzed and saw a bright yellow ball of light on the wall. This was the first of a series of unusual nightmares she experienced, some of which involved sleep paralysis and further unusual experiences. Dream images included people wearing "old-time" clothing, sexual situations involving priests and misshapen or diseased individuals, and images of huge penises. After a particularly unusual sleep paralysis event that was coupled with the sighting of a ball of light on the wall, Manny told her sister and mother of her experience. They decided that something paranormal was occurring because Nancy and Mrs. M. had had similar experiences, and both had seen an orange orb of light on their bedroom walls.

Family members wanted to see if they could observe the phenomena collectively. They stayed up all night in the room where the most events had occurred. This resulted in various unusual experiences. Manny tried to photograph the strange light orbs they witnessed. Although the lights appeared to be stimulated by her efforts to photograph them, the developed photographs revealed nothing unusual.

Other unusual events began occurring almost daily. Family members began to experience an itching, tingling, or burning sensation on their faces and ankles. They heard unexplained knocks, voices, whistling, music, and percussion instruments. They felt paralyzed and saw apparitions and anomalous lights. On various occasions, two family members had simultaneous experiences. Once Nancy felt paralyzed by a phenomenon, and Manny observed that Nancy and the air around her appeared orange and strangely misty. On another occasion, Manny and Mick simultaneously heard an apparitional distorted voice. They reported hearing a low, unintelligible voice like "a record being played at the wrong speed." But when they described their perceptions to each other and imitated the sound, they found that their imitations were different.

Mr. M. decided to investigate the phenomena by spending the night in the bed that was the focus of many incidents. He devised a plan, coordinated with Mick, in which he would indicate with his hand if he believed that anything unusual was occurring. He made a special effort to remain alert. During the night, a "funny feeling" swept over him, and he became paralyzed. He felt as if someone were physically pushing his feet, and he heard a knocking noise. He was unable to gesture to Mick, who was resting nearby on the floor. While Mick was watching, the side of the mattress under Mr. M.'s feet raised about eight to twelve inches. After it came down, Mr. M. was able to move and talk with his son.

The family was alarmed at this turn of events and asked Methodist, Pentecostal, and Catholic religious leaders to alleviate the disturbances. These leaders' prayers and ministrations had only temporary effects. The family also consulted psychics and fortune-tellers, hoping for advice. The information they obtained was not consistent or convincing. The apparitions and nightmares provided few clues regarding the source of the haunting beyond the images of people in nineteenth-century clothing.

My original hypothesis regarding psychopathology did not seem adequate. The family members seemed to be acting logically in the face of unusual perceptions. They were skeptical regarding the claims and information that psychic experts provided. On October 5, 1979, I spent the night in the room where much of the anomalous activity had been reported. Mick, Manny, and Nancy joined me in waiting for phenomena to occur. We all felt the itching, burning sensation that was thought to indicate the presence of the anomalous force. I had experienced this sensation on my first visit and assumed that it was the result of suggestion.

I fell asleep around 3:30 A.M. At 6:30 A.M., I awoke feeling tired. I saw a red oval of light, approximately one foot by two feet on the wall by a window. I noted that this light fit the family members' descriptions of the "ghost light" except that it was red rather than orange or yellow. I heard no sound, but assumed that the light was caused by the tail lights of an automobile on the street in front of the house. I listened closely for the sound of a car engine. I thought to myself that I would not accept the light as anomalous unless it became brighter or indicated in some way that it had a paranormal quality. I felt that it needed to communicate with my unvoiced thoughts in order to prove its authenticity. The light was visible for about thirty seconds, then it faded.

Immediately afterward, Mick began moaning as if he was having a nightmare, and a rap sounded within the wall near my face. I called out that something was there, hoping to wake everyone up so that they could witness whatever further events might occur. Mick said, "I was having a dream in which I was crawling over the floor toward your bed. It [the spirit] wanted you out of the bed. It felt you had to wake up, to get up."

I checked the window to determine the probability that the light I saw was from a car. A thick shade was tightly drawn. The light could not have come through any of the windows. I wondered if Mick had created the effect in some way, perhaps by using a flashlight. We went back to sleep.

At about 7:30 A.M., I woke feeling a vibration under my chest (I was sleeping on my stomach). The bed was moving up and down; the vibrations were similar to those produced by a minor earthquake. I looked about the room and saw no lamps or pictures swaying. The experience continued for about thirty seconds. My heart was beating rapidly with excitement, and I felt for my pulse, wondering if the vibration coincided with my heart beat. By the time I located my own pulse, the movement had stopped. I perceived my heart to be beating more rapidly than the vibration. I wondered if a device had been inserted in the mattress that could have caused the sensation (I did not consider this a likely possibility). I decided that it would be inappropriate for me to cut open the mattress because I was a guest at the house. I felt exhausted but elated. I had experienced phenomena similar to those described by the family members. I now understood their puzzlement.

The family members' explanations for the haunting differed. Mrs. M. was certain that the source was demonic. Mr. M. was uncertain but had come to the opinion that the haunting was real. Mick, Manny, and Nancy believed that spirits produced the effects. Mrs. M.'s sister, who had a hearing impairment, had not had an experience. Robert remained skeptical regarding all paranormal explanations. The two children reported no experiences, and the adults did not wish to talk about incidents in front of them.

Mrs. M. was afraid that my presence might stimulate what she believed were demonic spirits. I was allowed to visit during the day because she felt safer then. She also was concerned that I would think that the family had a mental disorder even though I assured them that this was not my evaluation.

I visited the house on a weekly basis. Eventually, all family members had experiences. Robert awoke paralyzed one night and perceived that his forehead was being gently stroked by a female apparition wearing a white hood. He was convinced the phenomenon was real. Mrs. M.'s sister was with the family when a loud banging noise sounded on the wall. She rose to open the front door and found no one outside. The noise was heard by all present, and no one could explain it. The children heard various apparitional voices and experienced a few other events. Although various family members attempted to personalize the phenomena (Mick called it "Billy, the ghost"), no one devised a convincing explanation regarding a specific deceased person. Eventually, experiences became more subjective (unusual dreams and apparitional voices) and less dynamic (less movement of objects and no bed-shaking episodes).

I could not determine whether the events were supernatural, psychological, or fraudulent. In a way, family members were evaluating their own grounded hypotheses. Those who had not previously considered themselves religious (Mick, Robert, Manny, and Nancy) now accepted, without doubt, the doctrine of life after death and the existence of spirits and souls. They reasoned that because they had experienced similar events independently, the cause did not involve crowd contagion. Nancy noted, "Although it has been truly inconvenient to have had all these things happen, in a way I feel blessed. Not many people have the kind of certainty that I do. You just can't doubt when you have experienced it so many times."

Family members reporting the greatest number of previous anomalous experiences (Mick, Manny, and Nancy) also described the highest frequency of haunting experiences. Robert, who had had no previous anomalous experiences, and who made the greatest effort to remain skeptical, claimed the least number of incidents of haunting phenomena. I hypothesized that haunting environments have the capacity to affect beliefs even among those, like Mr. M. and Robert, who were not previously believers.[30] These observations formed the basis for a grounded hypothesis that I would test in future cases: that a person's previous frequency of anomalous experiences was correlated with that person's frequency of haunting experiences, and that nonbelievers can experience haunting events that affect their belief. Although people lacking previous anomalous experiences may witness multiple haunting episodes, those with previous experience tend to perceive a far larger number. After one year, for example, Robert described four events, while Mick reported about a hundred. Because the propensity for anomalous experience seems to have a physiological basis, we would expect that some people during Paleolithic eras also experienced haunting episodes. These events could have caused those who were nonbelievers to accept the existence of spirits.

Between 1979 and 1992 I conducted over a dozen investigations of hauntings. I formed the hypothesis that particular people are more subject to these episodes and that these people are potential shamans. I helped people conduct rituals that coincided with their religious beliefs. I monitored the frequency of anomalous events before and after the rituals and determined that frequency and severity generally declined afterward. The relationship between ritual and frequency of haunting events was not verified scientifically since I had no control groups.

In 1992 I investigated poltergeist and haunting cases in Durham, North Carolina, with the assistance of Anne, a nurse who performed the role of a psychic. Anne sometimes provided police departments with "psychic" information about missing bodies, but was an agnostic with regard to supernatural claims. Anne helped me to investigate a "suicide haunting" case in Durham, N.C. John (twenty-one) had fathered a child and had moved in with the nineteen-year-old mother, Tammy, and her family. He had attempted suicide several years before and had been hospitalized and released. A few months before Anne and I encountered the case, John had lost his job, then killed himself with a shotgun. Tammy's brother and John's infant son were in the house at the time. They heard the shot and rushed into John's room to discover him dead.

Family members stated that the poltergeist events had probably started before John's suicide, but that they had become extremely active afterward. Objects moved about inexplicably, particularly objects related to John. The family watched in horror and amazement when an ornamental doll vibrated and gradually cracked before their eyes. They took photographs of the event, documenting the progressive changes, which culminated in the left eye falling out. The pattern of the disfigurement of the doll's head coincided with the pattern of the shotgun-blast damage to John's head.

The day before our initial joint interview, Anne visited the house. When she asked, "John, are you here?" she heard a loud inexplicable popping sound in response. During our visit Anne and I observed that the family was grieving. They were guilt ridden and in need of therapy. They had talked with social workers and were counting on time to heal their psychological wounds. John's son, a small child who was beginning to speak, appeared to be regularly seeing apparitional images of his father. The family was uncertain how to cope with this situation. They felt that the counselor they had talked with could not help them deal with their paranormal perceptions.

I spent the night at their house, sleeping in the most haunted area, but perceived no anomalous events. The family had heard that Anne and I had previously alleviated a haunting disturbance in their area. In that case I had aided Anne in gaining information by hypnotizing her. Tammy's family requested that we attempt a therapeutic ritual, seeking to contact John. We described our procedure as an experiment, making no claims that we would demonstrate paranormal phenomena. With the family present, I hypnotized Anne and suggested that she would be able to communicate with John. She said that she could hear him speak and could repeat his words. I instructed the family members to ask questions that they might have for John. Here was a typical exchange:

Tammy: "John, why did you kill yourself? Why did you leave us?"

John (speaking through Anne): "I didn't see any way out at the time. Now I realize it was a mistake. I'm so very sorry. Please forgive me."

Tammy (crying): "John, why did you leave me?"

John: "Tammy, I'm so very sorry. Please forgive me."

After various tearful exchanges, I had the family members discuss with each other how they wished to interpret the experiment. Some felt that Anne's utterances had characteristics indicating that John had actually been present. Others were unsure but had been overwhelmed by their emotions. They stated that the experiment had been beneficial because various issues could now be addressed more openly.

Our intervention seemed to have an effect. Minor anomalous incidents continued but the more robust paranormal events, which caused destruction of property, ceased. It appeared that our ceremony reduced the frequency of the anomalous experiences but did not terminate them completely.

Without controlled experimental studies, it is improper to draw firm conclusions regarding such therapeutic interventions. The apparent success of our ritual could be attributed to Anne's theatrical and counseling ability or to a natural tendency for poltergeist phenomena to decline over time. Although it was not a scientifically controlled field experiment, this intervention illustrates how shamanic ceremonies, designed to deal with grief, support belief in life after death. Shamans use similar types of trance performance all over the world. The shaman provides latent suggestions that are effective for hypnotizable people even when they are not in trance. I hypothesize that undesirable haunting experiences are like psychosomatic disorders in that they are treatable through psychological interventions, particularly hypnosis.

Haunting and poltergeist phenomena are at one end of a continuum regarding control of anomalous experiences. Haunting victims cannot control the ghosts they perceive. Shamans represent the other end of the continuum; they conjure apparitions, use extrasensory perception, and induce psychokinetic effects during staged ceremonies. They receive information that allows negotiation with spirits, and many participants claim to gain benefits. Lay people playing with Ouija boards or engaging in similar folk rituals are near the midpoint of the continuum.

My observations supported my previously formed grounded hypotheses. Although believers reported more haunting episodes than others, skeptics were not immune to experiences that affected their beliefs. Interventions designed to reduce the incidence of haunting events could be successful. These findings support the argument that Paleolithic people who experienced hauntings would have, as a result, created folk traditions in a manner equivalent to modern groups.

GROUPS ELICITING PSYCHOKINESIS

Psychical researchers have attempted to produce psychokinetic phenomena experimentally.[31] Groups meet regularly and sit in darkness around a table with their hands on its surface. These sitter groups sometimes find that the table moves anomalously in a manner that answers questions addressed to it. Apparitions, raps, and other psychokinetic effects may also occur, phenomena that some participants attribute to spirits. These experiments allow researchers to determine the conditions required to elicit anomalous perceptions. A form of folklore has emerged among the sitter groups: participants find that perceived psychokinesis is far more likely to occur among four or more people than among three or fewer. They claim that anomalous effects occur less frequently when they are closely monitored and more frequently in darkness. Rapport within the group facilitates effects.[32]

A sitter group, in semi-darkness, perceived that the table levitates. Note possible muscular exertion by one person's thumb. I have found that participants often demonstrate a form of dissociation (like shamans) in that they reveal no awareness of these "fraudulent" acts. Some feel that fraud causes others to believe, which triggers authentic phenomena. (photo by Dr. J. T. Richards/Fortean Picture Library)

I have been in sitter groups where many anomalous events have been perceived[33] and in others where nothing unusual has happened. Some table movements are such that it seems unlikely that subconscious muscular movements explain the effects. It is my impression that special people (who have had many previous anomalous perceptions) must participate in order for "normal" people to have unusual experiences. Some participants seem to reveal dissociated states, similar to those of shamans. They create effects perceived as anomalous by others yet reveal no awareness of the behavior (see figure above). Some believe that such subconscious "tricks" stimulate actual paranormal events. Perhaps the most important conclusion that can be drawn from these studies is that groups are able to generate extremely complex ideologies based on the messages that the spirits provide. They interpret these ideologies as coming directly from the spirits rather than from their own minds.

Some experiments shed light on the role of ideology in the creation of anomalous effects. A group in Toronto devised a lengthy story describing the tragic, but fictitious, life of "Philip." Philip was a British nobleman who did not intervene when his wife caused his lover to be burned at the stake as a witch. The group sought to communicate with Philip paranormally. After meeting for over a year, they began hearing rapping sounds that answered their questions. They observed anomalous table movements, some of which they captured on video. These experiments demonstrated that a group's ideology need not be true for unusual experiences to occur.[34] The implication is that the human mind has the power to generate anomalous experiences—spirits are not required.

Anthropologists have witnessed psychokinesis within shamanic contexts.[35] Native American shamans, for example, communicate with spirits during tent-shaking ceremonies. Within many traditions, the shaman is tied securely and placed in a small tent. The tent then begins shaking, and voices emerge. Observers write that the tent shakes more vigorously than appears to be possible by normal muscular exertion. Sometimes tent movements answer questions using a kind of code. The spirits might be heard speaking in unusual voices, different from the shaman's, in a manner thought to be beyond the limits of ventriloquism. Observers claim that the voices provide information that was unknown to the participants (the location of a lost person, for example). Tent-shaking ceremonies (which are assumed by outsiders to involve deception) occurred, with variations, among many tribes in the northern part of North America.

Native American and sitter group effects share many features: communication with spirits through psychokinesis, attempts to preclude the possibility of fraud, anomalous acquisition of information regarding distant events, special areas shielded from sight and set aside for psychokinetic activity, unusual shaking of objects in a manner thought to be beyond human ability, spiritual diagnosis and healing, the development of complex ideologies, and the assumption by skeptics that "paranormal" events are produced through sleight of hand. Accounts of similar forms of anomalous performance can be found within the anthropological literature regarding shamanism in societies all over the world.[36] The close equivalencies among cultures seem more than coincidental: shamans may use similar forms of conjuring without any formal training and without having had contact with others who use the same strategies. Because it is so difficult to explain how the diffusion of certain techniques has occurred, the "diffusion explanation" seems implausible. We might speculate that various performance commonalities are derived from similar sleight-of-hand techniques as well as effects derived from neurophysiological features involving dissociation.

Performances of the same type as those observed within sitter and shamanic groups have shaped religious history. The Spiritualist movement began in 1848 with poltergeist events among the Fox family in Hydesville, New York. Like sitter groups, the Fox children gained control over the phenomena. Two of the Fox children, Kate and Margaretta, developed a code for communicating with the source of the raps they were hearing, and in 1849 and 1850 they gave performances in major towns in the northeastern

United States. The rapping phenomenon spread so that by 1853 there were perhaps thirty thousand people who could perform such exhibitions.[37] The Spiritualists performances acquired other manifestations, including table tipping, trance mediumship, direct voice phenomena (ostensibly ventriloquism), and other forms of conjuring. Notable Spiritualist mediums Daniel Dunglas Home, William Stainton Moses, Leonora Piper, Eusapia Palladina, Gladys Osborne Leonard, Pearl Curran, the Schneider brothers, and Mina Crandon were investigated, some of them under seemingly controlled conditions, by psychical researchers, and many of the researchers were baffled by what they observed.[38]

William Swatos Jr. and Loftur Reimur Gissurarson review historical evidence from Iceland supporting the argument that anomalous group experiences impact religious traditions. Their findings portray the traditional social-scientific view of religion as a reflection of collective consciousness as invalid. At the beginning of the twentieth century, Icelandic Christianity had no spiritualist tradition, but many Icelandic people's beliefs were affected by the perceptions they experienced during spiritualist seances. Swatos and Gissurarson document group anomalous perceptions that were not part of Iceland's previous Christian tradition, but that had an important impact on that country's religious development.[39]

This evidence suggests that Paleolithic groups also found that collective rituals allowed them to generate anomalous experiences. Talented individuals may also have performed ostensibly extrasensory and psychokinetic feats that created belief in spirits. The "spirits" they contacted may have created the complex myths and ideologies required for a complete shamanic religion. Group anomalous experiences can generate relatively complex religious systems within a short period of time.

REVIEW OF GROUNDED HYPOTHESES

My participant observation has led to four grounded hypotheses: that people claiming frequent anomalous experiences have a greater probability of experiencing haunting events; that nonbelievers within haunted environments may become believers as a result of their experiences; that group processes can increase or reduce the incidence of anomalous experiences, affecting belief; and that group anomalous experiences can directly generate and perpetuate complex religious ideologies. These hypotheses, which support the ritual healing theory, are amenable to testing by others.

Anthropological Studies

Some anthropologists discuss the multidimensional nature of spiritual healing.[40] Spiritual healing within complex societies often takes place within small religious groups where participation allows a person to gain a new, healthier identity. Anthropologists have gathered evidence revealing why some healers are more effective than others and why certain clients respond better to particular healers.

Thomas Csordas presents two cases that illustrate a spiritual healer's effect on Western patients' illnesses.[41] His cases reveal the complex psychological processes associated with spiritual healing. Father Felix, a Pentecostal Catholic healer, holds private healing sessions in a counseling room in his monastery. He also conducts "healing of ancestry" masses in the homes of supplicants. Father Felix believes that people's problems often stem from evil spirits and the effects of previous generations. He uses prayer, healing services, and masses to alleviate these problems.

In one case, Margo, a twenty-seven-year-old woman, suffered from panic attacks and intense anxiety. Father Felix conducted three private sessions and one home mass for healing of ancestry. Although Margo found these treatments to be somewhat beneficial, her healing by Father Felix ended when a psychopharmacologist decided she should undergo electroconvulsive therapy. Father Felix departed for a long sabbatical before she could resume her sessions with him. Csordas portrays skilled spiritual healing not as a one-shot cure but as an incremental transformation. The life situation that generated Margo's problems could not be fixed with one treatment, but required an on going process, not unlike that associated with traditional psychological counseling. Effective spiritual healing is often associated with the client gaining a more positive identity.[42]

Twenty-five-year-old Ralph had been diagnosed with paranoid schizophrenia, obsessive-compulsive disorder, probable dysthymic disorder, some symptoms of agoraphobia, panic disorder, fear of heights, epilepsy, and asthma. During his first session when he closed his eyes, he perceived purple rings expanding concentrically in his visual field while feeling warmth emanating from Father Felix's hands. Ralph felt a benign presence for two days after this session. Although in subsequent sessions he also experienced heat and color, these sensations progressively declined in intensity, and Ralph came to view the therapeutic process as unsatisfactory. He felt that Father Felix overinterpreted and misunderstood his experiences. This seemed to weaken the impact of Father Felix's treatment.[43] A treatment that was originally thought to be transforming was not permanently integrated into Ralph's life, and, as a result, it had little therapeutic impact.

Csordas provides insights into how ritual healing can be "analyzed in terms of participants' predispositions, their experience of empowerment, and their experience of transformation."[44] His cases illustrate the incremental aspect of ritual healing. Spiritual healing typically entails an ongoing process of suggestion, acceptance of suggestion, and transformation associated with social relationships. Csordas criticizes models of spiritual healing, such as James Dow's,[45] that fail to specify how persuasion occurs and how healers create in their subjects a disposition to be healed.

Some healers seem better able than other healers to determine which symbolic strategies will be effective for a particular patient. These healers use a wide variety of strategies while others are confined to merely one or two techniques. Effective treatments fit the patient's illness history, emotional predispositions, and present suppositions. People suffering from chronic illnesses may journey from healer to healer seeking relief, and many gain no benefits. Some spontaneous healings occur outside of formal healing rituals. For example, one of my informants stated:

> I must have visited half a dozen different healers. I found them all to be worthless! But when you have cancer, you are so terrified—you will try anything. I put up with so much b*** s***. All they want is your money! So I guess I basically gave up. Then one evening I was reading my Bible when I felt something sweep over me—I believe it was the power of the Bible—and I knew I would be OK. I can't say that I was completely healed right then and there but I believe that my recovery began that evening and it caused my cancer to go into remission.

As argued by Csordas, healing is more complex than a mere manipulation of symbols. The effectiveness of any particular method varies greatly. In the example case, the respondent found that spiritual healers who were effective for others could not heal her. It seems that each client presents a unique psychological condition that can block a healing method from being effective. This barrier is like a lock that healers try to open with their methodological keys. If a healer provides the correct key, the one that is suitable for the client's lock, the person is healed. Shamans accept a particular methodological key as valid when it heals their own problems; then they use that method on others. They prosper if their collection of keys overcomes the psychological obstacles most common within their culture. The ritual healing theory portrays the process as physiologically based, yet embedded within the cultural context.

Surveys, folklore data, participant observation, and anthropological evidence provide insights into how anomalous experiences shape, and are constructed by, social interaction. The data support the ritual healing scenario. Cognitively open Paleolithic people in powerless positions would have been most vulnerable to psychosomatic, physical, and spiritual problems. These cognitively open people would have had a propensity for anomalous experience leading to belief in spirits, souls, life after death, and magical abilities. They would have benefited from ritual therapies and, as a result, some would have become the first shamans. Paleolithic groups probably found that collective experiences supported and extended their supernatural beliefs. They accepted the complex ideologies that their "spirits" devised for them. Shamanic healing originated in and worked among Paleolithic groups in a manner similar to the way it works among modern people.

Conclusion

Criticisms and Challenges

"How are you doing, Tom?" I asked the spine surgery patient. "Not very well," he replied, his face contorted. "I'm hurting and I'm not due for my pain pills until four P.M.—that's over an hour from now! They're trying to reach my doctor right now, but it's hard putting up with this pain."

I was visiting patients as part of a participant observation study at the Hershey Medical Center, Pennsylvania State College of Medicine. The purpose of the program was to uncover methods for making hospital treatment more humane.[1] "Many people find that they get relief through hypnosis," I told Tom. "Would you like to try it?"

"I'll try anything," Tom replied, "but I can hardly think straight with this pain, so I doubt I can be hypnotized. I'm not into weird stuff."

"Some people benefit more than others, but almost everyone gets some relief," I stated. "I merely provide suggestions, and you respond because it makes you feel better."

"I'll try anything!" Tom stated again.

"Well, to start off, I'm asking that you focus your attention on your feet. Let all your awareness be on your feet, and let your feet become completely relaxed." Tom looked at me while I talked, then let his eyelids fall closed. "You can feel the tension going out of your feet, and as I talk to you, you feel your feet becoming more and more relaxed. Your feet are becoming completely relaxed." I spoke in the rhythmic manner that causes hypnotic subjects to listen closely.

I then began suggesting that Tom relax his legs completely, saying that if he felt discomfort in his back as a result of his operation, he could allow that sensation to pass into the background of his mind, so that his discomfort would diminish as the exercise continued. I repeated these suggestions for other body parts—arms, chest, shoulders, neck, head—saying that he could relax the muscles in each area. I then suggested that he could see images very clearly in his mind's eye. I suggested that he imagine that he had arrived at a beautiful, still lake where the water was calm and peaceful and that his mind would become as calm and peaceful as the lake. I told him that the sun was shining down on him with a healing energy and that he felt a sense of peace and comfort as a result. He could feel the sun's healing rays helping him to become more and more relaxed . . . completely relaxed . . . completely at peace . . . completely in harmony with the beautiful scene he saw in his mind's eye. I remained silent for a while, allowing Tom to experience the tranquil images I had suggested. Tom was breathing quietly with his eyes closed. I watched his breathing change as he fell peacefully asleep.

People's reactions to this story illustrate the nature of scientific paradigms. Because my account does not mention supernatural forces, most people attribute Tom's analgesia to hypnotic suggestion. But some scholars argue that hypnosis is not real, that it entails merely expectation and role playing. Many of their arguments revolve around the definition of hypnosis and the inability of hypnosis researchers to describe exactly how hypnosis works. Lay people such as Tom who have experienced hypnosis find these arguments puzzling and somewhat silly. They know that hypnosis works.

The ritual healing theory is subject to similar criticisms. It posits that spiritual healing's effectiveness is correlated with hypnotizability. If I had prayed with Tom using repetitive, suggestive words, I could have achieved results equivalent to those gained through hypnosis. I might have shouted, "TOM, IN THE NAME OF JESUS, YOUR PAIN IS GONE!" and in the proper context my waking suggestion would be effective—his pain would disappear. Such commands might even increase the rate of physiological healing. For critics, this possibility merely compounds the problem: neither spiritual healing nor hypnosis is completely understood.

The folklore evidence makes the problem even more difficult: many accounts are anomalous—things happen that are beyond what can be explained by hypnosis. Suppose a healer in another building prayed for Tom and, as a result, his pain was alleviated. This story would be anomalous because we would not be able to use hypnosis theory to explain it. Or suppose Tom described passing out from the pain and seeing his deceased grandmother who told him, "Tom, I'm healing your back—you will be leaving the hospital this afternoon"—and the prophesy turned out to be true. This story would be even more implausible. We might visualize a continuum of stories: some would fit within scientific paradigms, but as we approached the other end of the spectrum, events would be increasingly difficult to explain.

Folklore collections portray this continuum: accounts vary in their degree of implausibility. Observers' evaluation of accounts that lie along this spectrum affects their thinking about the ritual healing theory. Some people reject the notion that hypnotic processes occur, and others believe in hypnosis. Some people attribute unusual effects to unknown mental processes, and others believe in supernatural powers. Some people reject all folklore evidence supporting the ritual healing theory because they cannot believe the more implausible stories. These positions are associated with different scientific paradigms. The ritual healing theory appeals to people who accept hypnosis as real, who suspect that mental processes produce extremely unusual experiences, yet believe that the supernatural has relatively little impact or is nonexistent. In contrast, criticism of the ritual healing theory generally arises from three sources: philosophical prejudice, definitional problems, and paradigm problems.

Philosophical Prejudice

Scholars often exhibit prejudice toward anomalous experiences. They argue that accounts of them are inappropriate for scientific study. Because anomalous reports sometimes involve deception, distortion of memory, and cognitive flaws, some scholars regard them as repulsive or trite. A prestigious reviewer of my book *Wondrous Events* stated that the publisher should "apply decent standards of selection and editing" so that those accounts that are "utter trash" would be left out. He wrote that such reports make the book a "lurid exploitation paperback of the kind sold to subnormals at the supermarket checkout."[2] Such comments imply that scientists should debunk occult beliefs and stigmatize believers as "subnormal." Yet if we decide that reports of contacts with the dead are so implausible that they should be dismissed, we will be forced to ignore the Christian resurrection story as well as other important accounts that are central to religious traditions. We cannot throw out certain stories as "utter trash" if we wish to understand why people believe as they do. Anomalous accounts are an inherent part of religion—all folk religions include belief in such events. The unexplained qualities within people's experiences are an inherent feature within folk religious belief, and they provide the foundation for shamanism, the first religion. If we throw out the stories of anomalous events, we cannot understand shamanism or other religions.

One problem is that, contrary to the desires of some skeptics, folklore stories *do* induce belief. As noted in chapter 5, many students stated that they gained an understanding of why people believe as they do after they interviewed them regarding their anomalous experiences. People have experiences, they tell others, and folk traditions result. If sociologists of religion wish to understand why people believe as they do, they must be willing to listen to people's stories.

That is not to say that we must accept all anomalous accounts as true. Critics are applying appropriate skepticism when they argue that stories should not be given equal consideration because "exceptional claims require exceptional proof." The philosopher David Hume provided a philosophical foundation for this argument: because a miracle is a violation of the laws of nature (which have been established by experience), no testimony is sufficient to establish a miracle unless it would be even more miraculous if

that testimony were a lie.[3] Hume's doctrine aids scientists in making deci-
sions regarding the publication of research in scientific "border" areas. Based
on this position, they reject studies advocating extremely anomalous claims
because no testimony would ever be sufficient to justify them.[4]

But the ritual healing theory does not seek to prove anomalous claims; it
merely acknowledges that people report such claims. It falls within the Darwinian paradigm and does not make exceptional claims beyond those of
modern evolutionists. Researchers can appraise the theory by testing hypotheses using standard scientific methodologies.

Definitional Problems

Some critics argue that key terms that are essential to the ritual healing
theory are so inadequately defined that hypotheses using these terms cannot be tested.[5] This criticism is sometimes linked with paradigm problems,
which I will discuss in a moment. But there is an extensive literature that
reviews the definitions of key terms: *religious sentiment*,[6] *altered states of consciousness*,[7] *shamanism*,[8] and *hypnosis*.[9] I cite this literature, provide standard definitions, and do not use these terms in any special sense. Definitional problems have not prevented previous researchers from presenting
theories and testing hypotheses using these terms.

Religious Sentiment   I begin with Anthony Wallace's definition of religion: "a set of rituals, rationalized by myth, which mobilized supernatural
powers for the purpose of achieving or preventing transformations of state
in man and nature"[10] This is the definition provided by the most widely
used anthropology textbook in the United States.[11] Webster defines *sentiment*
as "an attitude, thought, or judgment prompted by feeling; a predilection."

The ritual healing theory allows a better understanding of Wallace's definition. Shamanic rituals are repetitive actions that can stimulate therapeutic states of consciousness. Because modern primates use rituals, we can assume that hominids and later *Homo sapiens* also used them to generate
therapeutic transformations. The genotypes associated with these therapeutic benefits would have increased over the millennia, and the experiences
associated with these genotypes may have become the basis for all religious
traditions. Rituals are believed to mobilize supernatural powers because the
hypnotic processes by which they work are associated with anomalous experiences that generate belief in spirits, souls, life after death, and magical
abilities. Religion can be defined as the culturally prescribed set of behaviors that is devised as a result of the physiologically based propensity to respond to ritual suggestion. Religion involves human mind-brain relationships, something we are just beginning to understand. Within the ritual
healing paradigm, religion cannot be precisely defined because it involves
physiological processes that are not yet fully understood.

Altered States of Consciousness   I use Charles Tart's definition of altered states of consciousness, "a qualitative shift from normal patterns of mental functioning, a subjective sensation associated with behavioral changes,"[12] even though one critic argues, "ASC thus might include anything from sneezing to boozing, from enjoying an especially fine sunset to suffering gastroenteritis. . . . Such a notion is too broad and loose to be useful as an explanation."[13] Tart pioneered the study of other forms of consciousness with his book *Altered States of Consciousness* in 1969. His work called attention to the wide variety of forms of consciousness beyond waking and sleeping.

Allan Hobson's model of consciousness, which I reviewed in chapter 4, describes the physiological basis for conscious states. Anthropologists who observe altered states of consciousness in the field find it possible to investigate ASC even though consciousness is not completely understood. They have found that the forms of ASC associated with ritual involve increased suggestibility and unique physiological processes.[14] The ritual healing theory hypothesizes that specific physiological processes are associated with corresponding hypnotic, anomalous, and religious experiences. With research, further physiological parameters can be uncovered, and more precise explanations devised. Science never achieves complete certainty regarding any phenomenon but, through research, gains greater definitional precision over time.

Shamanism   A critic of my 1997 article "Shamanic Healing, Human Evolution and the Origin of Religion" writes: "Shamanism, another key term, is left undefined. The author does say that 'shamanic/hypnotic rituals' are 'acts which manipulate cultural symbols while inducing ASC.' However, because ASC is so broad and because symbols are omnipresent, such acts might include virtually anything that humans do." I argue, in response, that not all human acts manipulate symbols or induce altered states of consciousness.

Winkelman analyzes the ways in which anthropologists have defined *shamanism,* then offers his own empirical definition based on ethnographic studies. He defines *shamanism* as a practice that involves soul flight and ecstatic communication with the spirit world.[15] Analyzing his sample of ethnographic data, Winkelman found that healers in sedentary societies differ from those in hunting and gathering societies. He labels those who practice the altered form of shamanism in sedentary societies "shamanic healers." Most anthropologists recognize that shamanism, and religion in general, have changed over time as the social system has become more complex.[16]

Guthrie argues that shamanism includes malevolent magic that would place those hypnotizable people at risk.[17] If a society's shamans used malevolent magic more frequently than they used therapeutic magic, shamanic practices would eliminate genes associated with hypnotizability. But to my knowledge, no anthropologist who studies shamanism believes that the majority of shamans' efforts are malevolent. Healing is the primary shamanic activity.[18] I discuss mathematical analyses of cost-benefit ratios associated with the shamanic syndrome in the appendix.

Many anthropologists use the word *shamanism* within their writing. Many, and probably most, believe that altered states of consciousness were involved in Paleolithic shamanic practice.[19] Dickson writes:

> The ethnographic record . . . suggests that the religious life of Upper Paleolithic societies depended upon shaman, part-time practitioners who vigorously and directly sought to confront the spirit world in "ecstatic encounters." Given the high value that historic and recent food collectors place on achieving these special or altered states of consciousness, there no doubt was widespread group or communal participation in the religious life of the Upper Paleolithic.[20]

Dickson presents this position without defining the terms *shaman, altered states of consciousness,* or *religious* because these terms are understood by anthropologists. Demand for more precise definitions beyond those that are used by anthropologists seems a rhetorical ploy designed to defuse interest in the ritual healing theory.

Hypnosis   A critic of my 1997 article mocks the idea that hypnosis occurs among "unspecified ancient primates, in chimpanzees being groomed, and apparently even in animals feigning death." He criticizes the notion that "people may unconsciously hypnotize themselves" and states, "This uncertainty about hypnosis is especially deleterious to the argument. If hypnosis is to underlie religion in general, we need a convincing account both of its psychodynamics and of its presence in religion."[21]

The article cites ample supporting evidence, including Schumaker's summary of studies that link hypnotic processes with religion.[22] Over the past two hundred years hypnosis researchers have devised standardized tests that measure hypnotizability and have made great strides toward determining the physiological processes associated with the phenomenon.[23] The number of researchers who argue that *hypnosis* is insufficiently defined has decreased as researchers have identified physiological correlates of hypnotizability. The word *hypnotizability* is used as a term of convenience within the ritual healing theory to label a hypothetical collection of physiological structures that have been shaped by the use of ritual over the past few million years. We can gain a better understanding of these structures through further research, and the ritual healing theory provides a paradigm for such exploration.

Progress within a scientific field toward ever greater definitional precision is not unusual. When Darwin published his theory of evolution, he could not specify exactly what was being selected. Biologists during his era had little understanding of genetics and no knowledge of DNA. Present uncertainties do not constitute insurmountable obstacles. At present there is no universally agreed-upon definition of the term *gene,* for example, but this does not prevent advances in the field of genetics.[24]

Paradigm Problems

Thomas Kuhn argues that the bulk of conventional scientific research, or "normal science," consists of the production of expected solutions to pre-scribed problems according to standardized procedures. This sort of re-search is not oriented toward the pursuit of novel facts or theories but to-ward solving puzzles generated by existing theories. Kuhn refers to paradigms as "universally recognized scientific achievements that for a time provide model problems and solutions to a community of practitioners."[25] Kuhn's idea of the structure of scientific revolutions can be summarized as follows: A paradigm leads to normal science. Normal science discovers anomalies. Anomalies bring about crises, which, in turn, produce revolu-tion. New paradigms emerge from scientific revolutions.[26]

Much of social scientists' work is pursued within the parameters of specific paradigms. Prominent paradigms stress the importance of society and social-ization, and the cultural source theory portrays religious and anomalous ex-periences as cultural products.[27] Although many religious people think that their faith has experiential bases, the cultural source theory is accepted by the majority of social scientists.[28] A sophisticated version of the cultural source paradigm is articulated by critics who attack the ritual healing theory:

> The view (first associated with Schleiermacher and later with Otto) that reli-gion consists essentially of unmediated experience now is widespread in popu-lar culture. But it also has been convincingly called into question by Proudfoot (1985), who urges that what we do and undergo *becomes* experience only upon being interpreted. In Proudfoot's example, a woodsman, seeing a dark shape ahead on the trail, experiences fear when he interprets the shape as a bear. But when his companion persuasively reinterprets the shape as a log, the experience of fear changes to one of reassurance. Thus, experience comprises a sensation plus an interpretation. Since any given sensation can be interpreted in numerous ways, interpretation is key.[29]

Cultural source theorists regard religion as that nonexistent bear. Accord-ing to this theory, there is no specifically religious sentiment or source of religious belief—there are only interpretations of normal sensations that experiencers understand as religious.[30] "A person identifies an experience as religious when he comes to believe that the best explanation of what has happened to him is a religious one."[31] This argument is based on the sup-position that people cannot accurately interpret sensations derived from their own internal physiological processes but must depend on environ-mental cues and cultural guidance.

Wayne Proudfoot's argument is supported by an experiment conducted by Stanley Schachter and Jerome Singer, who hypothesized that emotions consist of a general, diffuse physiological arousal and a subsequent cognitive label.[32] They tested this hypothesis by injecting epinephrine (a hormone that produces heart palpitations, flushing, and tremors) into subjects in three groups. The first group was correctly informed of the drug's effects. The second group was incorrectly told that the drug would produce numbness and a slight headache. The third group was provided with no explanation. After receiving the injection, subjects were then exposed to one of two situations. Some of the subjects waited in a room with a euphoric confederate who tossed paper balls and played with a hula hoop, and some waited with a confederate who pretended to be angry at the experimenters. Schachter and Singer concluded that subjects who were not given a correct physiological explanation for their sensation labeled their physiological arousal in a manner that reflected their environment. Subjects in the euphoric environment labeled their epinephrine-induced sensation as euphoria, and subjects in the angry environment labeled their sensation as anger. Subjects who were provided with the correct explanation for their arousal were not influenced by the environmental cues.

Relying in part on the results of this study, Proudfoot argues that emotions are not self-apparent but are the product of environmental cues and culturally shaped interpretations. By implication, there is no such thing as a purely religious experience; people merely label sensations as religious because the label is appropriate within their cultural setting. Experiences, then, cannot occur prior to beliefs and concepts; instead, experiences are generated partly from beliefs and concepts. Hence, religious experiences, like the experience of being frightened by a nonexistent bear, never occur independently. They are shaped by belief or are produced by interpretations situated in belief.

Not everyone accepts Proudfoot's theory. Sallie King presents a coffee metaphor that she considers more appropriate than Proudfoot's metaphor of the nonexistent bear. People who drink coffee for the first time frame their descriptions of the experience within a cultural and linguistic context. Because of childhood memories, comments by others, and cultural traditions, the inexperienced drinkers are influenced by preconceptions. But like Schachter and Singer's subjects, the coffee drinkers have been exposed to a drug that produces a physiological effect, and that effect is constant regardless of the coffee drinkers' cultural context. "In the end, though drinking coffee is a mediated experience, that mediation is a relatively insignificant element of the experience itself."[33] King does not equate the relatively constant nature of the effects of caffeine with mystical experience, but her metaphor reveals the weakness of Proudfoot's suggestion that religious experiences with universal features are produced only through interpretation. Furthermore, both religious experiences and caffeine-induced experiences are sometimes at variance with the preconceptions and cultures within which they occur. "There are, after all, mystical innovators and founders of traditions. . . . For this to be possible something more, something new must come to the mystic above and beyond that for which the tradition with its teaching and training prepares him or her."[34]

Many scholars assume that all mystical experiences are "unformed" or "vague" (like a nonexistent bear) and that, as a result, interpretations of them are completely governed by culture. One problem may be that these scholars often base their opinions on the stories collected by William James and published in his *Varieties of Religious Experience* (1902).[35] They forget that most of James's account were from literary sources rather than social surveys. The types of anomalous experiences that people most frequently describe differ from the nebulous episodes that James presents. My collection indicates that the sorts of experience people tell each other (thus generating folk religious traditions) are not equivalent to the effects of a drug and are not vague mystical perceptions. One person sees his dead grandfather. Another perceives that she has left her body. Another has a vision of his wife at the moment of her death. Although the social environment shapes these experiencers' interpretations, their perceptions often generate the same conclusions: belief in spirits, souls, life after death, and magical abilities. These episodes are not always produced by belief; sometimes they produce belief. King's coffee metaphor seems more appropriate than that of Proudfoot's nonexistent bear.

Furthermore, the empirical basis for Proudfoot's theory and for the Schachter and Singer experiment does not hold up under scrutiny. Christina Maslach argues that Schachter and Singer's findings are "not, in fact, supported by the evidence. The initial data analysis reveals no significant difference in self-reported emotion between any of the unexplained arousal groups and the placebo control groups."[36] Gary Marshall and Philip Zimbardo note that Schachter and Singer base their conclusions on "several somewhat questionable post hoc internal analyses, such as classifying placebo subjects with no change in pulse as being 'aroused.'"[37] They note that an experiment designed to overcome Schachter and Singer's methodological problems found no support for the original conclusions. Subjects who received the epinephrine injection but were not provided with an explanation for the side effects did not differ from the control group in either behavior or emotional effect.

In a review of the twenty years of research generated by the Schachter and Singer experiment, Rainer Reisenzein found no compelling evidence supporting Schachter's propositions.[38] The only consistent finding supporting the original thesis is that an emotional state will be intensified by misattributed arousal from an irrelevant source. More recent studies indicate that, although the relationship between cognition and emotion is complex, physiological sensations are associated with specific emotions. Not only can emotion redirect or interfere with cognitive activity, but cognitive activity also affects emotion.[39]

Other experiments further weaken Proudfoot's position. Researchers have found that some reactions to perception happen before interpretation occurs.[40] In addition, studies indicate that people vary with regard to their awareness of internal processes. Privately self-aware individuals are less likely than those who are not self-aware to be fooled into believing that a placebo drug will alter their physiological state.[41] One reviewer writes: "Once it is clear that Proudfoot's basic model for understanding emotions is, at best, highly suspect, it becomes glaringly apparent that the conclusions which he derives from this fragile 'evidence' are unwarranted and misleading."[42] A reinterpretation of the Schachter and Singer experimental paradigm in light of more recent research would conclude that explanations of religious experiences are influenced by environmental and cultural factors but not fully produced by them.

Walter Pahnke conducted an experiment that presents a more appropriate model for explaining the universal features within religious experience.[43] He assigned twenty divinity-student subjects to experimental and control groups. The experimental group was administered thirty milligrams of psilocybin, and the control group received two hundred milligrams of nicotinic acid. Nicotinic acid is a vitamin that causes feelings of warmth and tingling of the skin but has no effect on the mind. After receiving these drugs, all subjects took part in a lengthy Good Friday religious service. The atmosphere was "broadly comparable to that achieved by tribes who use natural psychedelic substances in their religious ceremonies," because the particular content and procedure of the ceremony was "familiar and meaningful."[44] Afterward the subjects were extensively evaluated regarding their religious perceptions. Those who had received psilocybin scored significantly higher with regard to virtually all categories measuring mystical experience. Pahnke obtained equivalent results in a parallel experiment, and other studies using psychedelic drugs reached similar conclusions.[45] Interviews of experimental subjects twenty-four to twenty-seven years after the Good Friday service indicated that they unanimously felt that their psilocybin experiences had a genuinely mystical quality, "one of the highpoints of their spiritual life. . . . [On the other hand], most of the control subjects could barely remember even a few details of the service."[46]

Pahnke's experiments imply that physiological processes affect the incidence of mystical experience. Although the studies do not disprove the claim that environment and culture shape the interpretations of unusual perceptions, the experiments refute Proudfoot's assertion that interpretation of experience is based *only* on exterior and cultural cues. In parallel fashion, spontaneous anomalous experiences—such as apparitions, paranormal dreams, waking ESP, out-of-body perceptions, sleep paralysis—seemingly also involve physiological functions. The interpretation of these episodes also are not based *only* on exterior cues and cultural conditioning but sometimes result in new beliefs that prevail throughout the experiencer's life.

Maslach conducted a variation of Schachter and Singer's experiment that provides another vantage point for thinking about religious experience. Subjects in the test group were hypnotized and provided with the suggestion that the following reactions will occur whenever you see the word start. When you see the word start, your heart will beat faster, your breathing will increase, there will be a sinking feeling in your stomach, and your hands will get moist. You will feel all of these sensations as soon as you see the word start, and they will last until I say to you, "That's all for now." When I say, "That's all for now," you will return to your normal state and feel relaxed and good. However, when you see the word start and experience these reactions, you will not know why you are feeling the way you are, or remember my telling you anything about it.[47]

The symptoms cited in these instructions were chosen since they are parallel to those occurring after injection of epinephrine, making this study analogous to that of Schachter and Singer. Maslach used a learning task to introduce the arousal cue "start" in a nonobvious way. She measured hypnotized and control subjects' heart rate and galvanic skin response to verify that the cue did indeed stimulate physiological arousal. Injection of a drug was not required. During interviews following the experiment, subjects indicated that they were unaware of the true cause of their hypnotically induced arousal.

Maslach's data fail to support Schachter and Singer's findings: subjects were not swayed by the behavior of confederates, and "in all cases, subjects with unexplained arousal reported negative emotions, irrespective of the confederate's mood."[48] The study also demonstrated that drugs are not required to induce physiological response within the experimental setting. Hypnotic suggestion can create these effects.

Maslach's study is parallel to those of Aaronson, who found that hypnotic suggestions generated unexpected experiences.[49] Aaronson's subjects were hypnotized and were programmed to experience a pervasive perceptual change posthypnotically. After hypnosis, subjects were observed for two hours, taken for an automobile ride over a standard course, and asked to write an account of their day's experience. Some subjects who were exposed to suggestions involving increased depth of visual field or changes in temporal experience reported mystical or psychedelic experiences. One subject stated that he had been "transported into an experience of great beauty . . . each object and its placement seemed part of a Divine order." He "felt he could do no less than spend his life serving God."[50] In further experiments, Aaronson used posthypnotic suggestion to cause four experimental subjects to experience an egoless state of emptiness, and in other attempts, feelings of cosmic unity. At least three of the subjects reported eight of nine criteria for mystical experience following both the inductions for emptiness and cosmic unity. In contrast, control subjects described only two criteria.[51]

The studies of Pahnke, Maslach, and Aaronson imply that mystical and religious experiences are associated with physiological processes and that these processes can be triggered by hypnotic suggestion. Latent, symbolic triggers could occur both within rituals or everyday life. It is easy to induce a wide variety of posthypnotic responses beyond those attempted by Aaronson: thirst, paralysis, nonvolitional muscular movements, analgesia, amnesia, and hallucination. As mentioned in chapter 4, I have conducted dozens of simulated shamanic ceremonies supporting this argument. Within religious contexts, such behavior would be attributed to supernatural forces.

The links between religious/mystical experiences and hypnotic suggestion could be investigated empirically. Just as Maslach stimulated physiological response using hypnotic suggestion, researchers could use hypnotic suggestion to induce religious and mystical experiences. Highly hypnotizable subjects, provided proper suggestions, could experience effects equivalent to drugs such as psilocybin. My studies suggest that researchers could simulate many shamanic effects. We might assume that spontaneously occurring events occur in a similar fashion: certain brain functions allow religious, mystical, and anomalous perceptions to occur.

Although criticisms of the ritual healing theory point out problem areas, empirical research suggests that these weaknesses can be overcome. Through research, we can achieve an increasingly greater understanding of cognitive functioning, including the processes associated with hypnotic, religious, mystical, and anomalous experience.

In 1979, Daniel Moerman published an article entitled "Anthropology of Symbolic Healing." He proposed a unified view of the way that healing works, cited evidence indicating that symbolic healing is often effective, argued that physiological processes link symbols to health, and outlined empirical methods for studying the influence of spiritual healers. He suggested that "we might be able to modify our propensity to learn about other cultures, and begin to learn from them."[52] He was aware of the evolutionary implications of his argument, noting, "If we conclude that there are direct relationships between symbols and health, and if health is somehow reflected in fitness, relative reproductive success, we are on the verge of a theory of the evolution of the human symbolic capacity itself."[53] Comments following Moerman's article portrayed varied reactions: some heralded his contribution as "laudable," others described his position as "old hat" (a rehash of accepted ideas), and others attacked his assumptions as "debatable."

Although philosophical issues are difficult to resolve, many of Moerman's positions pertained to physiology. Since the publication of his article, much progress has been made regarding psychophysiological mechanisms governing brain functioning and consciousness, and many of Moerman's arguments have become almost axiomatic. The fields of mind-body medicine and psychoneuroimmunology have uncovered fundamental physiological processes linking emotion and health. I expect criticisms of *Wondrous Healing* to be parallel to those directed towards Moerman—reviewers will express a variety of opinions—but I also predict that empirically based researchers will find support for the various hypotheses that can be derived from my theory (see appendix). The ritual healing theory provides a unifying paradigm portraying how religion evolved from the relationship between the human capacity for symbolism and health. By devising and testing hypotheses derived from this theory, we gain a better understanding of what it means to be human.

Appendix

The ritual healing theory is amenable to mathematical modeling. Many studies indicate a positive correlation between hypnotizability (as measured by standardized tests) and the benefits derived from hypnosis, typically pain relief. Assuming a linear relationship, this can be expressed mathematically as

$$B = (Q)(X)$$

where B = benefits, Q = the quality of the therapeutic ritual, and X = hypnotizability

From an evolutionary perspective, benefits refer to increased ability to propagate genes. But the benefits derived from rituals are also associated with costs. Rituals require time that could be spent doing something else: prehistoric people who devoted most of their energy to rituals would reduce their capacity for survival and for propagating their genes. Hypnotizability is also associated with costs. People who are more hypnotizable suffer from higher rates of psychosomatic illness and other psychological problems. An evolutionary equation includes terms to reflect the costs of conducting too many rituals and of being too hypnotizable:

$$B = (QRX) - R^3 - X^3$$

where B = survival benefits, Q = quality of ritual,
R = the number of rituals performed during a specific time period,
and X = the genetic basis for religion or "religiousity/hypnotizability"

The final R and X terms are cubed to ensure that B will not increase infinitely as R and X increase. (This example is for illustrative purposes; many alternate models are possible.) This equation portrays how elements within the ritual healing theory may interact. People in a society that has low-quality rituals gain fewer benefits. Those who practice rituals excessively or who are extremely hypnotizable are penalized (the cost is set equal to the cube of these variables). R can be interpreted as behavioral religiosity (measured in terms of time spent engaging in rituals). Although the variable X might be termed hypnotizability and measured by standardized tests, we should remember that this is a label of convenience: X in this equation refers to a measure of the biologically based propensity to be religious.

The formula illustrates the evolutionary processes creating religious propensity. Although rituals practiced by pre-linguistic hominids were undoubtedly rudimentary, they must have Q values that were sufficiently high because otherwise the genotypes associated with X would not have survived (in the above equation, Q must be greater than two for B to have a positive value).

My simulated prelinguistic ritual demonstrated that even without language hominid rituals could have facilitated hypnotic processes. Chart 2 shows the relationship between B (benefits) and R (number of rituals per time period) for quality of level ten ($Q = 10$), for various values of X. A hominid lacking hypnotic capacity ($X = 0$) would gain no benefits from performing rituals while an $X = 1$ hominid would find that two rituals were optimal, allowing eleven benefits. Those with $X = 3$ could gain thirty-six benefits by performing three rituals, but $X = 4$ hominids would wish to perform four rituals but could gain only thirty-two benefits from them. A highly hypnotizable $X = 5$ might also perform four rituals but could only gain about eleven benefits from doing so. All these hominids might perform more than their optimal number of rituals, in which case, these extra rituals would cause higher costs, reducing the overall benefits. In this equation the exact number of rituals providing optimal benefits for each individual is $\sqrt{10X/3}$.

The curves in chart 2 appear as hills, with the highest point ($X = 3$, $R = 3$, $B = 36$) signifying the point at which certain hominids were maximizing their opportunities for survival. Any hominid not at the highest point would achieve less than maximum benefits from ritual performance and would be at a survival disadvantage. Most animal populations have probably

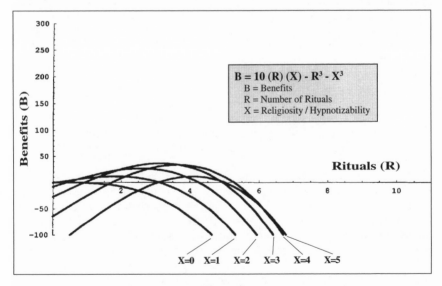

$$B = 10 \, (R) \, (X) - R^3 - X^3$$
B = Benefits
R = Number of Rituals
X = Religiosity / Hypnotizability

Chart 2

reached equilibrium with regard to the benefits they derive from their rituals, so unless some exterior factor comes into play, the situation is stable. The idea of evolutionary stability is central to the generally accepted concept of punctuated equilibrium.

But hominid evolution was not a static situation. During the past two million years, the increase in symbolizing ability led to rapidly spoken language. With language, rituals could be coupled with verbal suggestions, traditions, and ideologies, features that made them much more effective. The Q in the equation increased, causing those with higher X values to gain survival advantages. Chart 3 portrays the situation when $Q = 20$. An $X = 5$ human might wish to conduct six rituals and would gain 259 benefits, and an $X = 7$ human would gain 294 benefits by doing seven rituals. A highly hypnotizable $X = 9$ person, on the other hand, would gain only 199 benefits from eight rituals. The optimal way to gain ritual benefits in this society is to be an $X = 7$, $R = 7$ kind of person. Those whose values differ from these scores receive less benefits. Highly hypnotizable people would gain more from participating in rituals, but the costs of doing extra rituals and of being highly hypnotizable would be also be greater—they would suffer greater incidence of psychosomatic illnesses and spend more time engaging in religious activities as they seek to recover.

The curves in chart 3 look mountainlike compared to the benefit curves of the $Q = 10$ group. Humanity's increasing ability to use language resulted in a transformation: as ritual quality doubled, optimal X and R increased from three to seven, and the net benefits (after costs have been deducted) derived from religious practices by the person with optimal X and R values increased from 36 to 294. This model suggests that humans became religious as a result of language.

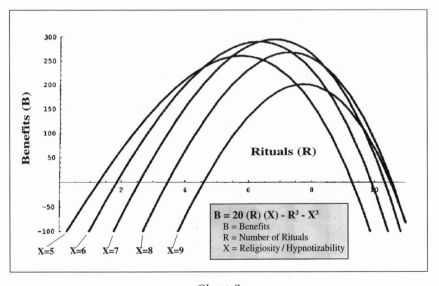

$$B = 20\ (R)\ (X) - R^3 - X^3$$
B = Benefits
R = Number of Rituals
X = Religiosity / Hypnotizability

Chart 3

This model also explains the variation in modern attitudes toward super-natural claims. People with low X values gain less from participating in rit-uals, and as a result many of them cannot understand why others are so re-ligious. They often assume that people who claim to have supernatural experiences are psychologically flawed in some manner. They are correct, in a way, because there are disadvantages associated with being highly reli-gious. People with high X scores feel the need to perform extra rituals, as they seek to restore themselves to physical and psychological well-being. Their religiosity is internal and deep, but they perceive that the earthly do-main does not reward this trait.

Both people with extremely high religiosity scores and those with ex-tremely low scores fulfill social functions within their group. Low-X, skepti-cal people help societies focus on practical affairs while high-X believers lead religious rituals that benefit those who desire and need a moderate level of ritual activity.

My equation is merely an illustrative model; its terms could be modified. Costs associated with X and R could be adjusted to determine the implica-tions of such changes. Let us suppose that the costs associated with per-forming rituals and with being religious were not as great as was assumed in the previous model. Suppose instead that

$$B = (QRX) - R^2 - X^2$$

Chart 4 portrays representations of this equation for various X's, assum-ing the prelinguistic hominid situation of $Q = 10$. The exact optimal number of rituals that each hominid would wish to perform to obtain optimal bene-fits is $10X/2$. The $X = 1$ hominid, for example, would wish to participate in 5

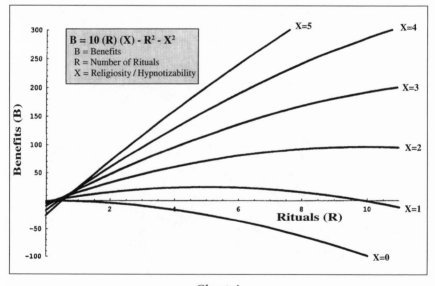

Chart 4

rituals. In evolutionary terms, there would be no limit to the religiosity that would develop within this model. An $X = 5$ hominid would wish to perform 25 rituals in order to obtain 600 benefits (off the scale of chart 4). In evolutionary terms, X would continually increase and, theoretically, an $X = 100$ hominid would have a survival advantage over all those with lower X values: this individual would wish to perform 500 rituals and would gain 240,000 benefits. According to this model, Q need not increase for evolutionary change to occur. According to this model, all animals continually evolve in a manner that leads to their performing more rituals regardless of the quality of those rituals. This model could be valid if we assume that evolutionary processes work slowly and that modern humans would benefit from achieving higher levels of religiosity than they have at present.

Chart 5 portrays lines representing X values for $B = 10\ RX - R - X$. Within this model, the penalties for performing rituals and for being religious are relatively low. Each hominid with religiosity greater than 0 would receive ever increasing benefits with increasing performance of rituals. This $B = 10$ $RX - R - X$ model does not represent what we observe in nature. Animals do not spend all their time performing rituals. We should conclude that models portraying higher penalties for rituals or religiosity are more valid.

In evaluating these models, we must consider the nature of evolutionary change. Some evolutionary theorists point out that processes that select for certain genotypes associated with a range of values would leave the population with little variance of those values. This would inhibit further evolutionary change. Other theorists argue that random genetic mutation would allow sustained evolutionary change.[1] Although this issue has not been completely resolved, exterior factors limit the growth of such values in most cases. The length of male peacocks' tail feathers provides an example.

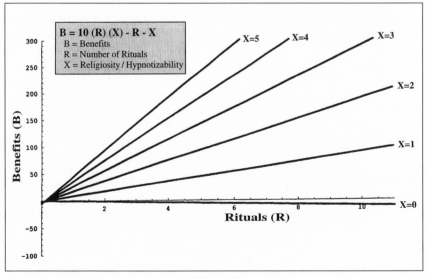

Chart 5

Although females are attracted to and mate with males with long tails, increasing the frequency of "long tail" genes, males with excessive plumage are unable to move about as easily. In the end, an optimal tail length results, reflecting a balance between attractiveness to females and physical agility.

Human religiosity probably has been limited in a similar manner; thus the $B = QRX - B^3 - X^3$ equation is probably superior to the $B = QRX - B^2 - X^2$ or $B = QRX - B - X$ equations because in the first equation the costs of X and R are sufficient to limit the continued increase of X. People who are more open to suggestion benefit from religious rituals but probably are at a disadvantage in other areas of human activity. Unfortunately, we do not have sufficient data to determine with certainty which equation most accurately portrays modern humans; if religiosity is still increasing gradually, the $B = QRX - B^2 - X^2$ equation could be the more valid.

These speculations allow insights into how evolutionary processes may have unfolded. These models portray how language could have been instrumental in changing the role of ritual within group life. In all models, increasing quality of rituals leads to greater religiosity. The fields of anthropology, neuroscience, and linguistics support a scenario in which speech evolved through a series of stages driven by natural selection.[2] These stages imply that the Q in all equation models gradually increased with the development of human language. This scenario, coinciding with the changing Q's in charts 2 and 3, requires spiritual healing to be only slightly effective; the evolutionary selection of genotypes related to religiosity would bring only small increases in X over time. By testing hypotheses derived from the ritual healing theory, we can refine the models and better define the concept of ritual quality. The ritual healing theory provides the following hypotheses.

First, paleoanthropologists will find artifacts, beyond those presently uncovered, that suggest that ancient humans benefited from rituals producing altered states of consciousness. Second, historians of medicine will find that all ancient medical texts link religion and medicine. The ritual healing theory suggests that all such texts imply that all ancient peoples used hypnotic and placebo healing systems.

Third, folklorists who collect accounts of spiritual healing will find that these narratives contain features indicating beneficial hypnotic and placebo processes. Fourth, large collections of anomalous experiences from any society will contain accounts of apparitions, waking ESP, paranormal dreams, psychokinesis, out-of-body experiences, sleep paralysis, and synchronistic events. Narrative elements within these accounts will contain cross-culturally consistent features. Analysis of these narratives will support the argument that these episodes have physiological bases. Respondents will report that these incidents have contributed to their belief in spirits, souls, life after death, and magical abilities—such ideologies increase the quality of therapeutic rituals. Fifth, case collections of apparitions, waking ESP, and paranormal dreams will reveal cross-culturally consistent structural features associated with family, death, temporal target, conviction, completeness of data, and severity of event. Apparition collections will portray universal "abnormal features of perception." These studies will imply that certain experiences have physiological bases.

Sixth, haunting cases, although shaped by their host society, will reveal cross-culturally consistent features. People with greater history of anomalous experience will tend to report greater incidence of haunting events. Haunting cases have the capacity to generate supernatural beliefs in previously skeptical people.

Seventh, social scientists can replicate studies that have revealed statistically significant correlations between hypnotizability, dissociative capacity, absorption, fantasy proneness, thinness of cognitive boundaries, and propensity for anomalous experience. Eighth, investigators will find that people who report frequent anomalous experiences tend to perform shamanic activities, engaging in spiritual healing. Ninth, psychoneuroimmunologists will continue discovering mechanisms that explain how psychological factors affect health. Their findings will provide insights into how rituals provide therapeutic benefits. Tenth, hypnosis researchers will generate further studies indicating the clinical value of hypnosis and the links between ritual, religion, and hypnotic experience. And eleventh, psychological studies will provide further insights into how rituals provide therapeutic benefits by affecting cognitive processes.

The realization that humans have a genetic propensity to benefit from ritual treatment has several potential practical applications. Anthropologists, psychologists, and sociologists could devise questionnaires for identifying people who would benefit from spiritual healing. These people could then be directed to practitioners who could provide the most benefit. The ritual healing theory predicts that hypnotizability, thinness of cognitive boundaries, and propensity for anomalous experience are important variables. People could evaluate their own capacity to respond to spiritual healers using self-administered questionnaires. Spiritual healing and self-hypnosis programs might be particularly effective for reducing postsurgical pain, chronic pain, and pain associated with terminal illness. Researchers could create video and audio programs providing instruction in self-hypnosis and including therapeutic suggestions specific to particular diseases and types of clients. Anthropologists and sociologists could monitor patients who have been treated by spiritual healers to determine the factors associated with therapeutic success.

Much evidence suggests that self-hypnosis training programs using video and Internet technologies would be cost effective. For many decades studies have indicated that psychological treatments can reduce postsurgical pain.[3] Clinical reports of success date back more than a century: English surgeon John Elliotson performed surgeries in the 1830s using hypnotism as the only means of anesthesia, and James Esdaile, a Scottish physician working in India, reported over three hundred major surgical cases in which hypnotism was the only anesthesia.[4] Ernest and Josephine Hilgard review thirty-two accounts of operations between 1955 and 1974 in which hypnotic analgesia was used in place of chemical analgesics or anesthetics.[5]

Major review papers summarize a quarter century of findings regarding psychological care of surgery patients: on average, surgical or coronary patients provided presurgical psychological support, particularly hypnotic suggestion, do better than patients who receive ordinary care; they stay in the hospital 2.37 fewer days than patients who do not receive such attention.[6] A meta-analysis of sixty-eight studies (with a total of 2,413 experimental subjects and 1,605 control group subjects) indicates that supportive preoperative instructions have favorable effects on operative outcomes.[7] R. Blankfield reviewed eighteen clinical trails in which suggestion, relaxation, or hypnosis was used in an attempt to facilitate recovery from surgery. Sixteen of the clinical studies found that intervention facilitated either physical or emotional recovery while two studies failed to document positive outcomes.[8] Other studies indicate that hypnotic suggestion can reduce blood loss during surgery[9] and requirements for chemical anesthesia.[10] Studies demonstrating a significant correlation between patient hypnotizability and benefits of hypnosis for surgery patients refute the argument that these are placebo effects.[11]

Psychological treatments can also be designed for mass audiences. A quantitative review of thirty-four controlled studies indicates that interventions that are unmatched in any way to the needs or coping styles of particular patients tend to be effective.[12] Standardized hypnotic suggestions are as effective as individualized suggestions,[13] and tape-recorded hypnotic inductions work equally well.[14] Studies indicate that self-training programs are particularly effective.[15] A recent well-controlled study demonstrates that standardized guidance in self-hypnotic relaxation significantly reduces pain and anxiety during invasive surgery.[16]

If self-hypnosis training programs were available via the Internet, patients could complete questionnaires to determine which program would be most suitable for someone with a particular problem, psychological propensity, and cultural background, then receive that program on-line. Tom, the patient described earlier, could log on to the Internet before his operation and take a questionnaire measuring his capacity for absorption and dissociation, and the thinness of his cognitive boundaries. Other questions would determine which programs coincide with his cultural background and expectations. He might be assigned to watch a firewalking demonstration followed by a discussion of the brain's ability to affect normally uncontrolled physiological parameters. A scientist might explain the benefits of self-hypnosis and guide him through practice sessions, teaching him how to use self-hypnosis to alleviate his postsurgical pain. If he were so inclined, he could listen to instructions from a spiritual healer representing a particular religion. He might join in prayers designed to reduce his pain and facilitate healing. He might even watch an on-line video of Wilasinee, the Thai shamanic healer described in the introduction of this book. Her pain- and heat-immunity performances could capture his attention, allowing her suggestions regarding healing to be effective. The cost of such a program would be low compared to the benefits derived.

By recognizing our innate capacity to respond to symbols, we can improve our health care systems. Rituals can heal, and modern technology allows healers to reach mass audiences efficiently.

Notes

1. McClenon (1994: 87–89).
2. Winkelman (1992: 48) operationally defines shamanism by noting traits of practitioners labeled as shamans in the anthropological literature. The magico-religious practitioners in his study, empirically clustered together and labeled as shamans, include the following: the !Kung Bushman *n/um kxoa-si;* the Samoyed *butode, 'dano, sawode,* and *tadibey;* the Semang *hala;* the Chukchee *ene nilit;* the Montagnais *manitousiou (okhi);* the Kaska *meta* and *nudita;* the Twana *bôswadas;* the Creek *alektca, hilishaya, owala,* and *kilas;* the Paiute *puha* and *puhaba;* the Cayua *pa?i* and *paye;* the Jivaro *wishinyu;* the Callinago *boyez;* and the Tupinamba *Pay* and *pagis.* Winkelman (2000: 66) summarizes the principal characteristics identifying shamanism as an epic phenomenon and reviews the methodology by which these characteristics were identified.
3. Winkelman (1992, 2000) makes this claim based on his analysis. All hunter and gatherer societies in his sample practiced shamanism, but no agricultural societies in his sample practiced pure forms of shamanism. He infers that shamanic religions were transformed into alternate forms with cultural change.
4. Eliade (1974); Heinze (1990).
5. Wickramasekera (1987: 12).
6. Spiegel and Albert (1983); Van Dyck and Hoogduin (1990).
7. Although I interviewed people who described surgical events that could not have been produced with mere sleight of hand, I did not see psychic surgeons perform such procedures. People's memories of events are often inaccurate.
8. Nadon and Kihlstrom (1987); Pekala, Kumar, and Cummings (1992); D. Richards (1990); Wagner and Ratzeburg (1987); Wickramasekera (1988); Wilson and Barber (1983).
9. Winkelman (1992, 2000).
10. Hobbes ([1651] 1991: 79).
11. Schleiermacher ([1799] 1996: 24).
12. Otto (1931).
13. Hay (1985).
14. Tylor ([1871] 1958).
15. Ibid., p. 29.
16. Spencer ([1876] 1969).
17. Müller (1889: 114).
18. B. Morris (1987).
19. Bellah (1964: 374).
20. Hufford (1982). The controversy over the "culture of disbelief" evokes so

much emotion that when I merely mentioned it in a journal article submission, one referee demanded that I prove that the culture exists.

21. Hufford (1982); McClenon (1984).

22. Guthrie (1993: 10).

23. Freud ([1927] 1964).

24. Ibid., p. 71.

25. Ibid., p. 23.

26. Ibid., p. 24.

27. Ibid., p. 47.

28. Malinowski ([1931] 1965: 111).

29. Ibid.

30. Greeley (1975).

31. Guthrie (1993); Koval (1990).

32. Durkheim ([1912] 1995: 214).

33. Ibid., p. 208.

34. Radcliffe-Brown (1922, [1939] 1965).

35. Harris (1997).

36. Kapferer (1983).

37. Runciman (1970).

38. Redekop (1967: 149).

39. Swatos and Gissurarson (1997).

40. Goldenweiser ([1917] 1965).

41. Malinowski ([1931] 1965: 111).

42. Evans-Pritchard (1965: 68).

43. Hay (1985). I have found that journal referees reject articles that point out this paradox. They inform me that Durkheim's theory cannot be directly tested in this fashion.

44. Durkheim ([1912] 1995: 208).

45. Wulff (1997: 49–116); Reynolds and Tanner (1995); Comfort (1979); Waller et al. (1990).

46. Trivers (1985) points out the problems with group-selection theories, but see Wilson and Sober (1994) for counterarguments.

47. Durkheim ([1912] 1995: 208).

48. Guthrie (1993: 5–6).

49. Ibid., p. 4.

50. Ibid., p. 7.

51. Ibid., p. 200.

52. Ibid., p. 7.

53. McClenon (2000).

54. Winkleman (2000); Schumaker (1990, 1995); Laughlin, McManus, and d'Aquili (1992); d'Aquili (1985); d'Aquili and Laughlin (1975); d'Aquili and Newberg (1993, 1998).

55. Guthrie (1993) reviews the major theorists to support this argument. Guthrie cites Evans-Pritchard (1965: 120–21); Geertz (1966: 1, 4); Wax (1984: 5); and Preus (1987: xvii).

56. Dennett (1995: 63).

1: The Evolution of Wondrous Healing

1. Bramwell (1959); Marcuse (1951, 1961).

2. Hoskovec and Svorad (1969).

3. Ibid.

4. Dingwall (1968).

5. Völgyesi (1966).

6. Lorenz (1966).

7. de Waal (1989).

8. Völgyesi (1966).

9. Rappaport (1979: 175) provides this definition but also argues that "no single feature of ritual is peculiar to it. It is in the conjunction of its features that it is unique." These features include formality (often stylized, repetitive, and stereotyped), performance, earnestness, and communication (pp. 175–79). The nature of human ritual coincides with, and facilitates, hypnotic processes.

10. d'Aquili (1985: 22).

11. Malan (1932: 314).

12. Reiser (1932).

13. Wulff (1997: 149–52).

14. Köhler (1927: 314–16).

15. Ibid., p. 315.

16. Goodall (1971, 1975).

17. Guthrie (1993).

18. Van Lawick-Goodall (1968: 273).

19. Ibid.; Goodall (1986).

20. Van Lawick-Goodall (1986: 400).

21. Ibid., p. 388.

22. Ibid., p. 391.

23. Ibid., p. 401.

24. Goodall, cited by Majno (1975: 12).

25. Ibid., p. 12.

26. Köhler (1927: 310).

27. Pert (1997).

28. Goodall (1986: 103).

29. Ibid.; Van Lawick-Goodall (1968).

30. Van Lawick-Goodall (1968: 273–274).

31. Needham (1967); Neher (1962); Lex (1979).

32. Benson, Beary, and Carol (1974).

33. Benson, Arns, and Hoffman (1981).

34. Ibid., p. 264.

35. Weiner (1992: 176).

36. Bohus, Koolhaas, and Korte (1991).

37. Haber and Barchas (1984).

38. Sapolsky (1990).

39. M. Fox (1974).

40. MacLean (1973).

41. Wright (1994).

42. Ibid., p. 26.

43. Lieberman (1984).

44. A. Walker (1993).

45. Lieberman (1984).

46. Corballis (1991: 98–105).

47. Van Lawick-Goodall (1968).

48. Darwin ([1871] 1952: 299).

49. Donald (1991: 168).

50. Lieberman (1984).

51. Donald (1991).

52. Vaneechoutte and Skoyles (1998).

53. Schumaker (1995).

54. Neher (1962).

55. Lieberman (1984: 322–23).

56. Ibid., p. 333.

57. White (1985).

58. Diamond (1989).

59. Blanc (1961: 124).

60. Ibid., p. 131.

61. Smirnov (1989).

62. Ibid., p. 221.

63. Davidson (1991).

64. Jolly and White (1995: 345).

65. Trinkaus (1983: 409–11).

66. Durham (1978).

67. Bayless (1970); *Fate Magazine* (1997); Gaddis (1970); Haynes (1972:31–59); Rogo (1982); Schul (1977); Sheldrake (1995: 13–19); Wylder (1978).

68. Seigel and Jarvik (1975).

69. Sigerist ([1951] 1987a: 112).

70. Pfeiffer (1982: 180).

71. Lewis-Williams and Dowson (1988).

72. Although I have observed shamanic practitioners painting images while in trance, this ability is rare (McClenon, 1994).

73. Lewin (1991).

74. Barabasz (1982); Barabasz and Barabasz (1989); Sanders and Reyher (1969); Wickramasekera (1977).

75. Hood (1995).

76. d'Aquili (1985: 22).

77. Ibid., p. 26.

78. Ibid., p. 22.

79. Ibid., p. 26.

80. Dawkins (1999).

81. Ramachandran and Blakeslee (1998).

82. C. Parker (1978).

83. Ibid.; Wilson and Sober (1994); Wright (1994).

84. d'Aquili (1985: 29).

85. Dickson (1990: 16–27).

86. Vialou (1998).

87. Dickson (1990: 215).

88. Harris (1997: 395).

89. Vialou (1998: 108–10).

90. Majno (1975: 29).

91. Sigerist ([1951] 1987a: 484).

92. Ibid., p. 412.

93. Sigerist ([1951] 1987a: 490); Majno (1975). Although some anthropologists

argue that ancient pharmacies contained many active and effective ingredients, these drugs, used within ritual situations, would also generate placebo and hypnotic results.

94. Sigerist ([1951] 1987a: 272).

95. Ibid., p. 286.

96. Sigerist ([1951] 1987a: 37).

97. Edelstein (1937); Prioreschi (1992).

98. Kouretas (1967a).

99. Kouretas (1967b).

100. Lewis-Williams and Dowson (1988); Nichols (1968).

101. Edelstein and Edelstein (1945); Sigerist ([1961] 1987b: 65).

102. Sigerist ([1961] 1987b: 68).

103. Ibid., p. 65.

104. Mac Hovec (1975, 1979); Stam and Spanos (1982).

105. Mac Hovec (1975, 1979).

106. Mac Hovec (1979: 88–89). Stam and Spanos (1982) critique this argument yet cite a huge body of literature supporting it. Their analysis focuses on the social factors influencing Asclepius-type healing.

107. Walker and Johnson (1974).

108. Belicki and Bowers (1982).

109. Gibson (1985).

110. Nadon and Kihlstrom (1987); Wilson and Barber (1983).

111. Walker and Johnson (1974: 369).

112. McClenon (1994).

113. Lumsden and Wilson (1983: 152).

114. Wickramasekera (1988).

115. Winkelman (1997) notes that 5 of the 11 hunter-gatherer shamans in his sample were coded for primary involvement in malevolent activities; 6 of the 11 were coded for secondary negative involvement. In general, anthropologists have observed that shamanism entails healing; negative activities are generally directed toward a distant target and should be less effective psychosomatically than healing activities directed toward a present target.

2: Fertility, Childbirth, and Suggestion

1. Dodds (1993).

2. Laderman (1987: 296).

3. Ibid.

4. Any large audience contains highly hypnotizable individuals who fall into trance when exposed to an induction. I have never found it difficult to locate subjects for demonstrations in the classroom or at professional conferences. Shamanic inductions are as effective as Western-style induction procedures.

5. Laderman (1987: 300).

6. Snow (1993: 213).

7. Harris (1997: 17).

8. Leavitt (1986).

9. The Progress of Nations 1996, League Table of Maternal Death. Available: <http://www.unicef.org/pon96/leag1wom.htm>.

10. Smirnov (1989).

11. McFalls (1979); Lobel (1994).

12. Although the effect of gender on psychic experience is unclear, women predominate in most case collections (Targ, Schlitz, and Irwin, 2000: 228–29).

13. Jemmott and Locke (1984).

14. Cebelin and Hirsch (1980); Syme (1975); Theorell and Rahe (1975).

15. Tennant (1988).

16. Henry and Stephens (1977).

17. Xiao and Ferin (1997).

18. Adams, Kaplan, and Koritnik (1985).

19. Shively, Laber-Laird, and Anton (1997).

20. McFalls (1979: 43) cites Ferreira (1965: 112); Kleegman and Kaufman (1966: 296); Noyes and Chapnick (1964: 547); Rutherford (1965: 114).

21. Utian, Goldfarb, and Rosenthal (1983: 241).

22. Belsey and Ware (1986).

23. McFalls (1979: 50).

24. Cabau and de Senarclens (1986).

25. Templeton and Penney (1982).

26. Southam (1960).

27. Cabau and de Senarclens (1986).

28. Noyes and Chapnick (1964).

29. Zigler-Shani et al. (1975).

30. Cabau and de Senarclens (1986).

31. Wilkinson et al. (1975: 245).

32. Zigler-Shani et al. (1975: 317).

33. Seibel and Taymor, 1982: 137).

34. Templeton and Penney (1982).

35. Cabau and de Senarclens (1986: 650).

36. Foldes (1975: 330).

37. Seibel and Taymor (1982).

38. McFalls (1979: 39, 40).

39. Utian, Goldfarb, and Rosenthal (1983: 241).

40. Boivin and Takefman (1995); Facchinetti et al. (1997).

41. Demyttenaere, Nijs, Steeno, Koninckx, and Evers-Kiebooms (1988); Demyttenaere, Bonte, Gheldof, Vervaeke, Meuleman, Vandershurem, and D'Hooghe (1998).

42. Masters and Johnson (1970: 264–65).

43. Kornhauser et al. (1975).

44. Crasilneck (1982: 52).

45. Brown and Chaves (1980).

46. Lambo (1974).

47. Rock (1986: 296–297).

48. Mann (1959: 709).

49. W. H. James (1963).

50. Tupper and Weil (1962, 1968).

51. Dick-Read (1933).

52. Gorsuch and Key (1974); McDonald (1968).

53. Kliment (1979).

54. Juznic, Vojvodic, and Avramovic (1979: 951).

55. Stahler, Stahler, and Gutanian et al. (1972).

56. Molinski (1975: 338).

57. Juznic, Vojvodic, and Avramovic (1979); Prill (1983).

58. Zichella et al. (1979: 731).

59. N. Morris (1983: 288).

60. Lobel (1994).

61. Wadhwa et al. (1996: 433).

62. Ibid.

63. Ibid., p. 432.

64. Harmon, Hynan, and Tyre (1990).

65. Ibid.; Jenkins and Pritchard (1993).

66. Omer, Palti, Friedlander (1986).

67. Zimmer et al. (1988).

68. Mehl (1994).

69. Barber (1984).

70. Snow (1993: 55).

71. Homer (1963: 367).

72. Dorson (1952: 153–54).

73. Bennett, Benson, and Kuiken (1986).

74. Disbrow, Bennett, and Owings (1993).

75. R. Hart (1980).

76. Rapkin, Straubing, and Holroyd (1991). Yet Hopkins, Jordan, and Lundy (1991) found that nonclinical hypnotized subjects were not able to control bleeding time.

77. Ewin (1984).

78. Swirsky-Sacchetti and Margolis (1986).

79. Dorson (1947: 108).

3: The Anthropology of Wondrous Healing

1. I describe many of these cases in my book *Wondrous Events* (1994).

2. Over one-fourth of Okinawan civilians were killed in the war. The social disruption was so great that many shamanic community ceremonies could no longer be performed. Shamanism there ceased to be a community-based religion and instead became a system in which clients contact practitioners.

3. This conversation took place before the publication of Thomas Csordas's *The Sacred Self* (1997), which discusses demonic possession in relation to self-concept.

4. Functionalists who attempt to explain situations without participating in them can arrive at silly conclusions. A functionalist observer might describe car drivers' visits to gas stations as rituals that are functional for the society and might argue that pumping gas generates social cohesiveness because ritual movements produce shared emotional experiences. The theorist might suggest that gas station rituals are required to generate the social cohesiveness required for societies with cars to survive. In parallel fashion, anthropologists watch rituals and assume these events create cohesion without actually "experiencing" the ritual effects.

5. Boddy (1994: 427).

6. Taussig (1987).

7. Karp (1989: 93).

8. Crapanzano and Garrison's (1977) anthology portrays "possessed" females gaining status.

9. Karp (1989: 93).

10. Young and Goulet (1994).

11. Turner (1992: 149).

12. Turner (1993: 9).

13. Although I think that the processes by which people perceive "spirit stuff" are not understood presently, this opinion is peripheral to the ritual healing theory.

14. Cardeña (1988).

15. Grindal (1983: 75).

16. Bergman (1973); Ortiz de Montellano (1975); Finkler (1985); Garrison (1977); Harner (1973); Kapferer (1983); Kleinman and Sung (1979); Kleinman (1980); Laderman (1987, 1991); Lambo (1974); Moerman (1979); Sharon (1978); Vogel (1970).

17. Kleinman and Sung (1979).

18. Curley (1973).

19. Edgerton (1980); Mullings (1984: 178).

20. Dodds (1993); Mullings (1984).

21. Glik (1986).

22. Csordas and Kleinman (1990); Finkler (1985); Frank (1973).

23. Hufford (1983, 1988).

24. Sammons (1992: 54).

25. Bass (2001); Lambo (1974); Warner (1985).

26. Luborsky (1975).

27. Christensen and Jacobson (1994).

28. Bowers and LeBaron (1983); Brown (1992).

29. Chapman, Goodell, and Wolff (1959); Ullman (1947).

30. Margolis et al. (1983).

31. Moore and Kaplan (1983).

32. Barber (1984).

33. Ibid.

34. Finkler (1985).

35. Kleinman (1980).

36. Tseng (1975).

37. Kleinman (1980: 369).

38. Kleinman and Sung (1979: 12).

39. Kapferer (1983: 59).

40. Horton (1967: 56).

41. Crapanzano (1973: 4–5).

42. Hufford (1983: 312, 313).

43. Cooper (1972); Dorson (1947); Emrich (1972); Freund and McGuire (1995); Hufford (1993); Snow (1993).

44. Bourguignon (1976: 18).

45. Kirsch (1990).

46. Benor (1990).

47. Ludwig (1966).

48. Mac Hovec (1975: 215).

49. Marcuse (1964).

50. Richeport (1992).

51. Mackett (1989).

52. Azuma and Stevenson (1988).

53. Donovan (1995).

54. Mac Hovec (1976).

55. Schumaker (1995).

56. Shostak (1981: 16).

57. Katz (1976, 1982).

58. Katz (1982: 123).

59. Katz (1976: 289).

60. Katz (1982: 238).

61. Wilson and Barber (1983); Morgan (1973).

62. Lewis (1971).

63. Hartmann (1991); McClenon (1994).

64. Katz (1982: 71).

65. Ibid., p. 122.

66. Honorton (1977); A. Parker (1975).

67. Honorton (1977).

68. Krippner and George (1986: 345); Krippner (1988).

69. Long (1977).

70. Rennie (forthcoming) discusses Eliade's opinions.

71. Katz (1982).

72. Kim (1967: 196).

73. Ibid.

74. Eliade (1974).

75. Thompson (1966); McClenon (1994: 115–16).

76. Coe (1958).

77. Walker (1977).

78. Leikind and McCarthy (1985).

79. Coe (1978).

80. Danforth (1989).

81. Harry Price (1936, 1937) reached similar conclusions in 1930. The fire-walker he observed showed special gymnastic abilities (he believed). Europeans attempting to imitate his accomplishment were burned.

82. Edgerton (1992).

83. Kane (1982).

84. Sammons (1992: 64).

85. Ewin (1984; 1986).

86. Margolis et al. (1983); Patterson et al. (1992).

87. Ullman (1947); Chapman, Goodell, and Wolff (1959).

88. Blake (1985).

89. Pekala and Ersek (1992/1993).

90. Hillig and Holroyd (1997/1998: 153).

91. Ibid., p. 161.

92. In McClenon (1994: 123–26) I describe previous experiments.

93. Rao and Palmer (1987); Broughton (1991).

94. Murdock (1945).

95. Bourguignon (1973); Eliade (1974).

96. Comaroff (1978).

97. Kleinman (1980: 372).

98. Dow (1986).

99. Csordas (1988).

100. The only way to distinguish hypnotic results from placebo effects is to demonstrate that the therapeutic outcome is correlated with hypnotizability. Placebo effects are not correlated with hypnotizability (as defined by standardized tests).

101. McClenon (1991, 1994).

102. Kleinman (1980: 372).

103. Bergman (2001: 172).

104. Skeptical observers have expressed alarm that these ceremonies induce false belief. In some cases, participants have stated that their beliefs have been strengthened. I stress that I am presenting a *simulated* shamanic ceremony. I argue that many "normal" religious ceremonies are far more effective for inducing anomalous experiences and instilling belief. Evaluation of paranormal claims is beyond the scope of my study.

105. Putnam (1988: 475).

106. McClenon (1994).

107. Kleinman and Sung (1979: 17).

108. Kapferer (1983).

109. Csordas (1988); Csordas and Kleinman (1990); Csordas (1997); Garrison (1977).

110. Csordas (1997).

111. Csordas (1988).

4: Wondrous Healing, Hypnotizability, and Folklore

1. Mehl-Madrona (1997: 32–33) and quoted by Krippner and Achterberg (2000: 353–54).

2. Kalweit (1992); Hultkrantz (1992); McClenon (1994: 222).

3. Nace et al. (1982).

4. Mehl-Madrona (1997: 29).

5. Baars (1997).

6. Hobson (1994).

7. Heide, Wadlington, and Lundy (1980).

8. Argyle and Beit-Hallahmi (1975: 97–99); Gibbons and de Jarnette (1972); Hood (1973); Nadon and Kihlstrom (1987); Pekala, Kumar, and Cummings (1992); Richards (1990); Schumaker (1995); Wagner and Ratzeburg (1987); Wickramasekera (1988); Wilson and Barber (1983).

9. Neppe (1983); Ross and Persinger (1987); Persinger (1984a, 1984b); Persinger and Valliant (1985).

10. Mandell (1980).

11. d'Aquili and Laughlin (1975); Mandell (1980).

12. Schumaker (1990: 16).

13. Persinger (1984a).

14. Dewhurst and Beard (1970).

15. Bear and Fedio (1977).

16. Ramachandran et al. (1997); Ramachandran and Blakeslee (1998).

17. Makarec and Persinger (1990); Ross and Persinger (1987); Persinger (1984b); Persinger and Valliant (1985).

18. Persinger and Makarec (1993: 33).

19. Ross and Persinger (1987).

20. Spiegel and Spiegel (1987).

21. Richards and Persinger (1991).

22. Persinger and Valliant (1985).

23. Wilson and Barber (1978); quote is from Persinger and Makarec (1993: 34).

24. Bear (1979).

25. Rossi (1986).

26. Tiller and Persinger (1994).

27. Hartmann (1991).

28. People with thin cognitive boundaries would be predicted to more frequently experience the Charles Bonnet syndrome: the mind fills in blind spots and scotomas with constructed images, which can include apparitions.

29. Winkelman (1992: 112).

30. Winkelman (2000).

31. Cardeña (1988).

32. Crawford and Gruzelier (1992).

33. Crawford (1994).

34. Crawford, Brown, and Moon (1993).

35. Hilgard (1977).

36. Morgan, Hilgard, and Davert (1970); Morgan (1973).

37. Rawlings (1972).

38. Duke (1969).

39. Dingwall (1968).

40. Rogo (1975: 43).

41. Bem and Honorton (1994); Honorton (1977); Targ, Schlitz, and Irwin (2000).

42. Schechter (1984)

43. Perry, Nadon, and Button (1992).

44. Shor and Orne (1962).

45. These first seven suggestions and the tenth suggestion are termed *nonvolitional* tasks.

46. This command tests for "inability to communicate."

47. Hilgard (1977: 115).

48. Sheehan (1992: 388).

49. Kok (1989) tested Singapore participants in a religious ritual involving self-inflicted wounds. Kok's study indicates that results of standardized hypnosis tests may be culturally specific. Although most subjects did not reveal high hypnotizability, those who did score high on the standardized test perceived their experience during the ritual as equivalent to hypnosis.

50. Morgan and Hilgard (1973); Piccione, Hilgard, and Zimbardo (1989).

51. Bernstein and Putnam (1986); Carlson and Putnam (1993).

52. Kirsch and Council (1992); Shames and Bowers (1992).

53. Nadon and Kihlstrom (1987); Pekala, Kumar, and Cummings (1992); Richards (1990); Wagner and Ratzeburg (1987); Wickramasekera (1988); Wilson and Barber (1983).

54. Tellegen and Atkinson (1974).

55. Schumaker (1990, 1995).

56. Weitzenhoffer and Sjoberg (1961); Hilgard and Tart (1966).

57. Woody, Bowers, and Oakman (1992: 25).

58. Bowers and LeBaron (1986: 464).

59. Schumaker (1995).

60. Ibid., p. 64.

61. Benor (1990); Murphy (1992: 257–83).

62. Sammons (1992); Schumaker (1995).

63. Bowers and LeBaron (1986); Brown (1992).

64. McClenon (1994).

65. The use of prayer cloths was deemed not to constitute a HEALRIT ritual because it does not involve repetitious activity.

5: The Seeds of Religion

1. Hufford (1982).

2. Although Proudfoot's 1985 book exemplifies an aspect of the cultural source theory, anonymous reviewers argue that, even if his arguments are true, a biological explanation could also be true. Hufford (1982) contends that one element within the cultural source position is a tendency to deem anomalous experience unworthy of discussion, too trivial to be of interest. Proudfoot's (1985) work is frequently cited to justify the position that anomalous experiences are not important.

3. Studies finding that these experiences are common include Glock and Stark (1965); Back and Bourque (1970); Greeley (1975, 1987); Hay (1990); Hay and Morisy (1978); McClenon (1994); Haraldsson and Houtkooper (1991).

4. Hay and Morisy (1978); Greeley (1975).

5. Hardy (1970, 1979).

6. Hay and Morisy (1978); Emmons and Sobal (1981); J. Fox (1992).

7. Hay (1979).

8. Greeley (1975).

9. Haraldsson and Houtkooper (1991); McClenon (1984).

10. Thomas and Cooper (1980). Of the contributors to my collection of 1,446 narratives from northeastern North Carolina not one reported "the feeling of being close to a powerful spiritual force that seemed to lift you out of yourself." I have found that this question elicits positive response from people who remember a wide variety of unusual events, including being "born again," out-of-body experiences, and other anomalous events that seem to be only vaguely related to mysticism.

11. Hufford (1985).

12. Greeley (1975, 1987).

13. Haraldsson and Houtkooper (1991).

14. McClenon (1994).

15. We should not assume that the percentage of a student group reporting a particular experience coincides with the percentage of people in the general population who would report the same experience. Alvarado (2000: 185) provides evidence indicating that 25% of students in 49 studies reported out-of-body experiences compared to only 10% of respondents in 5 studies of general populations.

16. These narratives are taken from McClenon (1994).

17. Emmons (1982); McClenon (1994); Rhine (1981); Schouten (1979, 1981, 1982); Virtanen (1990).

18. Sidgwick and Committee (1894).

19. Tyrrell ([1942] 1963: 25–26).

20. D. K. West (1948).

21. D. J. West (1990).

22. Gallup and Newport (1991).

23. J. Palmer (1979).

24. Hobbes ([1651] 1991: 79).

25. Emmons (1982: 75–92).

26. McClenon (2000).

27. Ramachandran and Blakeslee (1998).

28. One reviewer revealed a strange form of prejudice by demanding that I verify that all family members' experiences were *exactly* equivalent. My goal is not to verify paranormal claims but to analyze the stories that people tell.

29. Sidgwick and Committee (1894).

30. Ibid.

31. H. Hart (1959).

32. Virtanen (1990).

33. Broughton (1991); Irwin (1994).

34. Kurtz (1985).

35. Gurney, Myers, and Podmore ([1886] 1970).

36. Schouten (1979).

37. Schouten (1981).

38. Rhine (1981).

39. Schouten (1982).

40. Virtanen (1990).

41. Rhine (1981); Schouten (1979, 1981, 1982); Virtanen (1990).

42. Rhine (1954, 1981); Schouten (1982: 149–150); Virtanen (1990: 143).

43. Rhine (1951, 1981); Schouten (1979: 437–438; 1982: 147); Virtanen (1990: 132–36).

44. Rhine (1981: 130–32); Schouten (1979: 441; 1981: 42; 1982: 147).

45. Schouten (1979: 441; 1981: 42; 1982: 147).

46. Moody (1975).

47. Sabom (1982).

48. Ring (1980).

49. Zaleski (1987: 153).

50. Ibid., p. 7.

51. M. King (1985: 92).

52. Ibid., p. 93–94 and cited by Kellehear (1996: 31).

53. Kellehear, 1996: 40, 41).

54. McClenon (1991, 1994).

55. Guthrie (1993: 189).

56. Hufford (1982).

6: When the Seeds of Religion Sprout

1. Worrall (1970: 85).

2. Ibid.

3. Ibid.

4. J. Palmer (1979: 233–234).

5. Greeley (1975, 1987).

6. *Frequent experience* is defined as at least three different types of experience or "several times" of two types of experience or "many times" of one type of experience (McClenon, 1994: 30).

7. Wilson and Barber (1983).

8. Nadon and Kihlstrom (1987).

9. Dixon and Laurence (1992).

10. Pekala, Kumar, and Cummings (1992); Richards (1990); Targ, Schlitz, and Irwin (2000); Wagner and Ratzeburg (1987); Wickramasekera (1988).

11. McClenon (1994); Targ, Schlitz, and Irwin (2000).

12. Thalbourne and Delin (1994, 1999); Thalbourne et al. (1997).

13. Thalbourne and Delin (1999).

14. Hartmann (1991).

15. Bohus et al. (1991); Sapolsky (1990).

16. Gallagher (1995); Cockerham (1996).

17. Ayto (1990: 275–76).

18. Gauld and Cornell (1979).

19. Ibid.

20. During four semesters I asked students to seek out rootlore accounts because I regarded these stories as particularly interesting. For this reason, the number of rootlore narratives collected does not reflect the actual prevalence of these experiences.

21. Some Ouija board narratives specifically exclude the explanation that subconscious muscular movements produce the effect. Storytellers state that the marker moves without anyone touching it. Other accounts describe apparitional events coinciding with psychokinetic phenomena. Although these events seemingly exceed scientific explanation, they might be attributed to fabrication, false memory, or trickery.

22. McClenon (1994).

23. Emmons (1982); McClenon (1994).

24. Anne and many other respondents describe their experiences as part of a video presentation (Edwards and McClenon, 1993).

25. Eliade (1974).

26. Glaser and Strauss (1967).

27. Gallup and Newport (1991).

28. This case is described in McClenon (1994: 61–64).

29. Names of informants are pseudonyms.

20. My experiences in Baltimore did not cause me to regard myself as a believer. It was only after more compelling experiences during another investigation that I came to believe that similar anomalous episodes occur all over the world and are the basis for folk traditions involving the supernatural (McClenon, 1994).

31. Batcheldor (1966, 1979, 1984); Brookes-Smith (1973, 1975); Brookes-Smith and Hunt (1970); Owen and Sparrow (1976).

32. Batcheldor (1984); Richards (1982).

33. McClenon (1994).

34. Owen and Sparrow (1976).

35. Kalweit (1992); Hultkrantz (1992); Young and Goulet (1994).

36. Long (1977); Winkelman (1982).

37. Gauld (1968).

38. McClenon (1994).

39. Swatos and Gissurarson (1997).

40. Csordas (1997) reviews this literature.

41. Csordas (1988).

42. Csordas (1997).

43. Csordas (1988: 131).

44. Ibid., p. 132.

45. Dow (1986).

Conclusion: Criticisms and Challenges

1. McClenon (1996).

2. Bainbridge (1995).

3. Hume ([1748] 1988).

4. McClenon (1984).

5. Guthrie (1997).

6. Guthrie (1993).

7. Tart (1969); Winkelman (2000).

8. Winkelman (2000).

9. Fromm and Nash (1992); Orne (1977).

10. Wallace (1966: 107).

11. Haviland (1997).

12. Tart (1969).

13. Guthrie (1997: 356).

14. Winkelman (1986, 1992).

15. Winkelman (1992: 48).

16. Bellah (1964); Harris (1997); Wallace (1966).

17. Guthrie (1997: 356, 357).

18. Winkelman (1997).

19. Dickson (1990); Lewis-Williams and Dowson (1988).

20. Dickson (1990: 190–91).

21. Guthrie (1997: 356).

22. McClenon (1997a); Schumaker (1995).

23. Fromm and Nash (1992) review contemporary research. The field supports various peer-reviewed journals (for example, *International Journal of Clinical and Experimental Hypnosis, American Journal of Clinical Hypnosis*), and articles involving hypnosis often appear in the mainstream psychological journals.

24. Dawkins (1999: 28).

25. Kuhn (1970: viii).

26. One referee seemed to misunderstand my discussion of Kuhn (1970). He argued that my research should not be published because Kuhn's use of the term *anomalous* does not pertain to anomalous experiences. He argued that I should work within present paradigms until they are found to be invalid (I suppose he considers my use of the Darwinian paradigm unacceptable). This type of argumentation illustrates how the insistence on adherence to established paradigms prevents innovation.

27. Hufford (1982).

28. Some critics seem unable to conceive of paradigms outside the cultural source theory. They demand that the ritual healing theory explain phenomena such as power relationships and social conflict, issues within the domain of the cultural source paradigm.

29. Guthrie (1997: 356).

30. Most critical reviewers present few arguments when rejecting the ritual healing theory but merely cite Proudfoot (1985). After all, if there are no true religious experiences, then experiences cannot have led to religious beliefs.

31. Proudfoot (1985: 108).

32. Schachter and Singer (1962).

33. King (1988: 265).

34. Ibid., p. 267.

35. W. James ([1902] 1985).

36. Maslach (1979: 954).

37. Marshall and Zimbardo (1979: 981).

38. Reisenzein (1983).

39. Lazarus, Kanner, and Folkman (1980).

40. Ramachandran and Blakeslee (1998).

41. Gibbons et al. (1979).

42. Barnard (1992: 235).

43. Pahnke (1966).

44. Pahnke and Richards (1966: 192).

45. Pahnke (1967); Wulff (1997: 192).

46. Doblin (1991: 13).

47. Maslach (1979: 958).

48. Ibid., p. 963.

49. Aaronson (1966, 1967, 1968, 1969, 1970).

50. Aaronson (1968: 16–17).

51. Aaronson (1966).

52. Moerman (1979: 66).

53. Ibid., p. 77.

Appendix

1. Dawkins (1999).

2. Lieberman (1997).

3. Barber and Adrian (1982); McCaffery and Beebe (1989); Bonica (1990); Lawlis et al. (1985).

4. Esdaile ([1846] 1957).

5. Hilgard and Hilgard (1983).

6. Blankfield (1991); Mumford, Schlesinger, and Glass (1982); Rogers and Reich (1986).

7. Hathaway (1986).

8. Blankfield (1991).

9. Fredericks (1978) and Marmer (1959).

10. Bennett, Benson, and Kuiken (1986); R. Hart (1980).

11. Greenleaf et al. (1992); Rapkin, Straubing, and Holroyd (1991); Disbrow, Bennett, and Owings (1993).

12. Mumford, Schlesinger, and Glass (1982).

13. Van der Does and Van Dyck (1989).

14. R. Hart (1980).

15. Daake and Gueldner (1989); Syrjala, Cummings, and Donaldson (1992).

16. Lang et al. (2000).

References

Aaronson, Bernard S. 1966. "Behavior and the Place Names of Time." *American Journal of Hypnosis: Clinical, Experimental, Theoretical* 9: 1–17.

———. 1967. "Mystical and Schizophreniform States and the Experience of Depth." *Journal for the Scientific Study of Religion* 6: 246–52.

———. 1968. "Hypnotic Alterations of Space and Time." *International Journal of Parapsychology* 10: 5–36.

———. 1969. "Hypnosis, Depth Perception, and the Psychedelic Experience." Pp. 315–23 in *Altered States of Consciousness*, ed. C. T. Tart. New York: John Wiley and Sons.

———. 1970. "Some Hypnotic Analogues to the Psychedelic State." Pp. 279–95 in *Psychedelics: The Uses and Implications of Hallucinogenic Drugs*, ed. Bernard Aaronson and Humphrey Osmond. Garden City, N.Y.: Anchor.

Adams, M. R., J. R. Kaplan, and D. R. Koritnik. 1985. "Psychosocial Influences on Ovarian Endocrine and Ovulatory Function in *Macaca Fascicularis*." *Physiology and Behavior* 35: 935–40.

Alvarado, Carlos S. 2000. "Out-of-Body Experiences." Pp. 183–218 in *Varieties of Anomalous Experience: Examining the Scientific Evidence*, ed. Etzel Cardeña, Steven Jay Lynn, and Stanley Krippner. Washington, D.C.: American Psychological Association.

Argyle, Michael, and Benjamin Beit-Hallahmi. 1975. *The Social Psychology of Religion*. London: Routledge and Kegan Paul.

Ayto, John. 1990. *Dictionary of Word Origins*. New York: Arcade.

Azuma, Nagato, and Ian Stevenson. 1988. "'Psychic Surgery' in the Philippines as a Form of Group Hypnosis." *American Journal of Clinical Hypnosis* 31: 61–67.

Baars, Bernard J. 1997. *In the Theater of Consciousness: The Workplace of the Mind*. New York: Oxford University Press.

Back, Kurt W., and Linda Brookover Bourque. 1970. "Can Feelings Be Enumerated?" *Behavioral Science* 15: 487–96.

Bainbridge, William Sims. 1995. "Review of *Wondrous Events: Foundations of Religious Belief*." *Journal for the Scientific Study of Religion* 34: 533.

Barabasz, Arreed F. 1982. "Restricted Environmental Stimulation and the Enhancement of Hypnotizability: Pain, EEG Alpha, Skin Conductance and Temperature Responses." *International Journal of Clinical and Experimental Hypnosis* 30: 147–66.

Barabasz, Arreed F., and Marianne Barabasz. 1989. "Effects of Restricted Environmental Stimulation: Enhancement of Hypnotizability for Experimental and Chronic Pain Control." *International Journal of Clinical and Experimental Hypnosis* 37: 217–31.

Barber, J., and C. Adrian, eds. 1982. *Psychological Approaches to the Management of Pain*. New York: Brunner/Mazal.

Barber, T. X. 1984. "Changing 'Unchangeable' Bodily Processes by (Hypnotic) Suggestions: A New Look at Hypnosis, Cognitions, Imagining and the Mind-Body Problem." Pp. 69–128 in *Imagination and Healing*, ed. A. A. Sheikh. Farmingdale, N.Y.: Baywood.

Barnard, G. William. 1992. "Explaining the Unexplainable: Wayne Proudfoot's *Religious Experience*." *Journal of the American Academy of Religion* 60: 231–56.

Bass, Thomas. 2001. "Traditional African Psychotherapy: An Interview with Thomas Adeoye Lambo." Pp. 181–86 in *Magic, Witchcraft, and Religion: An Anthropological Study of the Supernatural*, ed. Arthur C. Lehmann and James E. Myers. Mountain View, Calif.: Mayfield.

Batcheldor, Kenneth J. 1966. "Report on the Case of Table Levitation and Associated Phenomena." *Journal of the Society for Psychical Research* 43: 339–59.

———. 1979. "PK in Sitter Groups." *Psychoenergetic Systems* 3: 77–93.

———. 1984. "Contributions to the Theory of PK Induction from Sitter-Group Work." *Journal of the American Society for Psychical Research* 78: 105–32.

Bayless, Raymond. 1970. *Animal Ghosts*. New York: University.

Bear, David M. 1979. "Temporal Lobe Epilepsy: A Syndrome of Sensory-Limbic Hyperconnectionism." *Cortex* 15: 357–84.

Bear, D., and P. Fedio. 1977. "Quantitative Analysis of Interictal Behavior in Temporal Lobe Epilepsy." *Archive of Neurology* 34: 454–67.

Belicki, Kathryn, and Patricia Bowers. 1982. "The Role of Demand Characteristics and Hypnotic Ability in Dream Change Following a Presleep Instruction." *Journal of Abnormal Psychology* 91: 426–32.

Bellah, Robert N. 1964. "Religious Evolution." *American Sociological Review* 29: 358–74.

Belsey, M. A., and Helen Ware, 1986. "Epidemiological, Social and Psychological Aspects of Infertility." Pp. 631–47 in *Infertility: Male and Female*, ed. V. Insler and B. Lunenfelf. New York: Churchill Livingstone.

Bem, Daryl J., and Charles Honorton. 1994. "Does Psi Exist? Replicable Evidence for an Anomalous Process of Information Transfer." *Psychological Bulletin* 115: 4–18.

Bennett, H. L., D. R. Benson, and D. A. Kuiken. 1986. "Preoperative Instruction for Decreased Bleeding during Spine Surgery." *Anesthesiology* 65: A245.

Benor, Daniel J. 1990. "Survey of Spiritual Healing Research." *Complementary Medical Research* 4: 9–33.

Benson, Herbert, Patricia A. Arns, and John W. Hoffman. 1981. "The Relaxation Response and Hypnosis." *International Journal of Clinical and Experimental Hypnosis* 29: 259–70.

Benson, Herbert, J. F. Beary, and M. P. Carol. 1974. "The Relaxation Response." *Psychiatry* 37: 37–46.

Bergman, Robert L. 1973. "A School for Medicine Men." *American Journal of Psychiatry* 130: 663–66.

———. 2001. "A School for Medicine Men." Pp. 168–72 in *Magic, Witchcraft, and Religion: An Anthropological Study of the Supernatural*, ed. Arthur C. Lehmann and James E. Myers. Mountain View, Calif.: Mayfield.

Bernstein, Eve, and Frank Putnam. 1986. "Development, Reliability, and Validity of a Dissociation Scale." *Journal of Nervous and Mental Disease* 174: 727–35.

Blake, Julianne. 1985. "Attribution of Power and the Transformation of Fear: An Empirical Study of Firewalking." *Psi Research* 4: 64–90.

Blanc, Alberto C. 1961. "Some Evidence for the Ideologies of Early Man." Pp. 119–36 in *Social Life of Early Man*, ed. Sherwood L. Washburn. Chicago: Aldine.

Blankfield, R. 1991. "Suggestion, Relaxation, and Hypnosis as Adjuncts in the Care of Surgery Patients: A Review of the Literature." *American Journal of Clinical Hypnosis* 33, no. 3: 172–87.

Boddy, Janice. 1994. "Spirit Possession Revisited: Beyond Instrumentality." *Annual Review of Anthropology* 23: 407–34.

Bohus, B., J. M. Koolhaas, and S. M. Korte. 1991. "Psychosocial Stress, Anxiety and Depression: Physiological and Neuroendocrine Correlates in Animal Models." Pp. 120–38 in *Stress and Related Disorders: From Adaptation to Dysfunction*, ed. A. R. Genazzani, G. Nappi, F. Petraglia, and E. Martignoni. Park Ridge, N.J.: The Parthenon Publishing Group.

Boivin, J. T., and J. E. Takefman. 1995. "Stress Level Across Stages of In Vitro Fertilization in Subsequently Pregnant Women." *Fertility and Sterility* 64, no. 4: 802–10.

Bonica, John J. 1990. *The Management of Pain*. Philadelphia: Lippincott Williams & Wilkins.

Bourguignon, Erika. 1973. *Religion, Altered States of Consciousness, and Social Change.* Columbus: Ohio State University Press.

———. 1976. "The Effectiveness of Religious Healing Movements: A Review of the Literature." *Transcultural Psychiatric Research Review* 13: 5–21.

Bowers, Kenneth S., and Samuel LeBaron. 1986. "Hypnosis and Hypnotizability: Implications for Clinical Intervention." *Hospital and Community Psychiatry* 37: 457–67.

Bramwell, J. M. 1959. *Hypnotism: Its History, Practice and Theory.* New York: Julian.

Brookes-Smith, Colin. 1973. "Data-Tape Recorded Experimental PK Phenomena." *Journal of the Society for Psychical Research* 47: 69–89.

———. 1975. "Paranormal Electrical Conductance Phenomena." *Journal of the Society for Psychical Research* 48: 73–86.

Brookes-Smith, Colin, and D. W. Hunt. 1970. "Some Experiments in Psychokinesis." *Journal of the Society for Psychical Research* 45: 265–81.

Broughton, Richard S. 1991. *Parapsychology: The Controversial Science.* New York: Ballantine.

Brown, Daniel P. 1992. "Clinical Hypnosis Research since 1986." Pp. 427–58 in *Contemporary Hypnosis Research*, ed. Erika Fromm and Michael R. Nash. New York: Guilford.

Brown, Jude M., and John F. Chaves. 1980. "Hypnosis in the Treatment of Sexual Dysfunction." *Journal of Sex and Marital Therapy* 6: 63–74.

Cabau, Anne, and Myriam de Senarclens. 1986. "Psychological Aspects of Infertility." Pp. 648–72 in *Infertility: Male and Female*, ed. Vaclav Insler and Bruno Lunenfeld. Edinburgh: Churchill Livingstone.

Cardeña, Etzel. 1988. "Deep Hypnosis and Shamanism: Convergences and Divergences." Pp. 289–303 in *Proceedings of the Fourth International Conference on the Study of Shamanism and Alternate Modes of Healing*, ed. Ruth-Inge Heinze. Madison, Wisc.: A-R Editions.

Carlson, Eve Bernstein, and Frank W. Putnam. 1993. "An Update on the Dissociative Experiences Scale." *Dissociation* 6: 16–27.

Cebelin, M. S., and C. S. Hirsch. 1980. "Human Stress Cardiomyopathy: Myocardial Lesions in Victims of Homicidal Assaults without Internal Injuries." *Human Pathology* 11: 123–32.

Chapman L., H. Goodell, and H. Wolff. 1959. "Increased Inflammatory Reaction Induced by Central Nervous System Activity." *Transactions of the Association of American Physicians* 72: 84–110.

Christensen, Andrew, and Neil S. Jacobson. 1994. "Who (or What) Can Do Psychotherapy: The Status and Challenge of Nonprofessional Therapies." *Psychological Science* 5: 8–14.

Cockerham, William C. 1996. *Sociology of Mental Disorder,* 4th ed. Upper Saddle River, N.J.: Prentice Hall.

Coe, Mayne Reid, Jr. 1958. Reprint. "Firewalking and Related Behaviors." *Journal of the American Society for Psychical Research* 52: 85–97. Originally in *The Psychological Record* 7 (1957): 101–10.

———. 1978. "Safely Across the Fiery Pit." *Fate Magazine* (June): 84–86.

Comaroff, John. 1978. "Medicine and Culture: Some Anthropological Perspectives." *Social Science and Medicine* 12 (B): 247–54.

Comfort, Alex. 1979. *I and That: Notes on the Biology of Religion.* New York: Crown.

Cooper, Horton. 1972. *North Carolina Mountain Folklore and Miscellany.* Murfreesboro, N.C.: Johnson.

Corballis, Michael C. 1991. *The Lopsided Ape: The Evolution of the Generative Mind.* New York: Oxford University Press.

Crapanzano, Vincent. 1973. *The Hamadsha: A Study in Moroccan Ethnopsychiatry.* Berkeley, Calif.: University of California Press.

Crapanzano, Vincent, and Vivian Garrison. 1977. *Case Studies in Spirit Possession.* New York: John Wiley and Sons.

Crasilneck, Harold B. 1982. "A Follow-up Study in the Use of Hypnotherapy in the Treatment of Psychogenic Impotency." *American Journal of Clinical Hypnosis* 25: 52–61.

Crawford, Helen J. 1994. "Brain Dynamics and Hypnosis: Attentional and Disattentional Processes." *International Journal of Clinical and Experimental Hypnosis* 42: 204–32.

Crawford, Helen J., Audrey M. Brown, and Charles E. Moon. 1993. "Sustained Attentional and Disattentional Abilities: Differences Between Low and Highly Hypnotizable Persons." *Journal of Abnormal Psychology* 102: 534–43.

Crawford, Helen J., and John H. Gruzelier. 1992. "A Midstream View of the Neuropsychophysiology of Hypnosis: Recent Research and Future Directions." Pp. 227–66 in *Contemporary Hypnosis Research,* ed. Erika Fromm and Michael R. Nash. New York: Guilford.

Crossan, John Dominic. 1991. *The Historical Jesus: The Life of a Mediterranean Jewish Peasant.* San Francisco: HarperSanFrancisco.

Csordas, Thomas J. 1988. "Elements of Charismatic Persuasion and Healing." *Medical Anthropology Quarterly* 2: 120–42.

———. 1997. *The Sacred Self: A Cultural Phenomenology of Charismatic Healing.* San Francisco: University of California Press.

Csordas, Thomas J., and Arthur Kleinman. 1990. "The Therapeutic Process." Pp. 11–25 in *Medical Anthropology: A Handbook of Theory and Method,* ed. Thomas M. Johnson and Carolyn F. Sargent, 11–25. Westport, Conn.: Greenwood.

Curley, Richard T. 1973. *Elders, Shades, and Women: Ceremonial Change in Lango, Uganda.* Berkeley, Calif.: University of California Press.

Daake, D. R., and S. H. Gueldner. 1989. "Imagery Instruction and the Control of Postsurgical Pain." *Applied Nursing Research* 2: 114–20.

Danforth, Loring M. 1989. *Firewalking and Religious Healing: The Anastenaria of Greece and the American Firewalking Movement.* Princeton, N.J.: Princeton University Press.

d'Aquili, Eugene G. 1985. "Human Ceremonial Ritual and the Modulation of Aggression." *Zygon: Journal of Religion and Science* 20: 21–30.

d'Aquili, Eugene G., and Charles Laughlin, Jr. 1975. "The Biopsychological Determinants of Religious Ritual Behavior." *Zygon: Journal of Religion and Science* 10: 32–38.

d'Aquili, Eugene G., and Andrew B. Newberg. 1993. "Religious and Mystical States: A Neuropsychological Model." *Zygon: Journal of Religion and Science* 28: 177–99.

———. 1998. "The Neuropsychological Basis of Religions, or Why God Won't Go Away." *Zygon: Journal of Religion and Science* 33: 187–201.

Darwin, Charles. [1871] 1952. *The Origin of Species by Means of Natural Selection/The Descent of Man and Selection in Relation to Sex*. Chicago: Encyclopaedia Britannica. Original edition, *Descent of Man*, London: John Murray.

Davidson, Ian. 1991. "The Archaeology of Language Origins: A Review." *Antiquity* 65: 39–48.

Dawkins, Richard. 1999. Reprint. *The Selfish Gene*. Oxford: Oxford University Press. Original edition, 1976.

Dean, Douglas, and E. Brame. 1975. "Physical Changes in Water by the Laying-On of Hands." *Proceedings of the Second International Congress of Psychotronic Research*. Paris: Institute Metaphysique International.

Demyttenaere, K., L. Bonte, M. Gheldof, M. Vervaeke, C. Meuleman, D. Vanderschuerem, T. D'Hooghe. 1998. "Coping Style and Depression Level Influence Outcome in In Vitro Fertilization." *Fertility and Sterility* 69, no. 6: 1026–33.

Demyttenaere, K., P. Nijs, O. Steeno, P. Koninckx, G. Evers-Kiebooms. 1988. "Anxiety and Conception Rates in Donor Insemination." *Journal of Psychosomatic Obstetrics and Gynecology* 8: 175–81.

Dennett, Daniel C. 1995. *Darwin's Dangerous Idea: Evolution and the Meanings of Life*. New York: Simon and Schuster.

de Waal, Frans. 1989. *Peacemaking among Primates*. Cambridge: Harvard University Press.

Dewhurst, Kenneth, and A. W. Beard. 1970. "Sudden Religious Conversions in Temporal Lobe Epilepsy." *British Journal of Psychiatry* 117: 497–507.

Diamond, Jared M. 1989. "Were Neanderthals the First Humans to Bury Their Dead?" *Nature* 340 (August 3): 344.

Dick-Read, Grantly. 1933. *Natural Childbirth*. London: Heinemann.

Dickson, D. Bruce. 1990. *The Dawn of Belief: Religion in the Upper Paleolithic of Southwestern Europe*. Tucson, Ariz.: University of Arizona Press.

Dingwall, Eric J., ed. 1968. *Abnormal Hypnotic Phenomena*. 4 vols. London: Churchill.

Disbrow, E., H. Bennett, and J. Owings. 1993. "Preoperative Instructions for Decreased Bleeding during Spine Surgery." *Anesthesiology* 65: 245.

Dixon, Michael, and Jean-Roch Laurence. 1992. "Two Hundred Years of Hypnosis Research: Questions Resolved? Questions Unanswered!" Pp. 34–66 in *Contemporary Hypnosis Research*, ed. Erika Fromm and Michael R. Nash. New York: Guilford.

Doblin, Rick. 1991. "Pahnke's 'Good Friday Experiment': A Long-Term Follow-Up and Methodological Critique." *Journal of Transpersonal Psychology* 23, no. 1: 1–28. Available: <http://www.druglibrary.org/schaffer/lsd/doblin.htm>

Dodds, J. Scott. 1993. *Healers of Ghana*. Princeton, N.J.: Films for the Humanities and Sciences. Videocassette.

Donald, Merlin. 1991. *Origins of the Modern Mind: Three Stages in the Evolution of Culture and Cognition*. Cambridge: Harvard University Press.

Donovan, James. 1995. "Multiple Personality, Hypnosis, and Possession Trance." Pp. 99–112 in *6th Yearbook of Cross-Cultural Medicine and Psychotherapy 1994*, ed. R. Quekelbherge and D. Eigner. Berlin: VWB, Verlag für Wissenschaft und Bildung.

Dorson, Richard M. 1947. "Blood Stoppers." *Southern Folklore Quarterly* 11, no. 2: 105–18.

———. 1952. *Bloodstoppers and Bearwalkers: Folk Traditions of the Upper Peninsula*. Cambridge: Harvard University Press.

Dow, James. 1986. "Universal Aspects of Symbolic Healing: A Theoretical Synthesis." *American Anthropologist* 88: 56–69.

Duke, J. D. 1969. "Relatedness and Waking Suggestibility." *International Journal of Clinical and Experimental Hypnosis* 17: 242–50.

Durham, William H. 1978. "The Coevolution of Human Biology and Culture." Pp. 11–32 in *Human Behavior and Adaptation*, ed. N. Blurton Jones and V. Reynolds. New York: Halsted.

Durkheim, Émile. [1912] 1995. *The Elementary Forms of Religious Life*. Translated by Karen E. Fields. New York: Free Press, Simon and Schuster. Original edition, *Les Formes élémentaires de la vie religieuse: Le système totémique en Australie*, Paris: F. Alcan.

Edelstein, Emma J., and Ludwig Edelstein. 1945. *Asclepius: A Collection and Interpretation of the Testimonies*. 2 vols. Baltimore: Johns Hopkins Press.

Edelstein, Ludwig. 1937. "Greek Medicine in Its Relations to Religion and Magic." *Bulletin of the Institute of the History of Medicine* 5: 201–46.

Edgerton, Robert B. 1980. "Traditional Treatment for Mental Illness in Africa: A Review." *Culture, Medicine and Psychiatry* 4: 167–89.

———. 1992. *Sick Societies: Challenging the Myth of Primitive Harmony*. New York: Free Press.

Edwards, Emily, and James McClenon. 1993. *Wondrous Events: Foundations of Folk Belief*. University Park: Pennsylvania State University. Videocassette.

Eliade, Mircea. 1974. Reprint. *Shamanism: Archaic Techniques of Ecstasy*. Translated by Willard R. Trask. Princeton, N.J.: Princeton University Press. Original English edition, 1966.

Emmons, Charles F. 1982. *Chinese Ghosts and ESP: A Study of Paranormal Beliefs and Experiences*. Metuchen, N.J.: Scarecrow Press.

Emmons, Charles F., and J. Sobal. 1981. "Paranormal Beliefs: Testing the Marginality Hypothesis." *Sociological Focus* 14: 49–56.

Emrich, Duncan. 1972. *Folklore on the American Land*. Boston: Little, Brown and Company.

Esdaile, J. [1846] 1957. *Hypnosis in Medicine and Surgery*. Reprint, New York: Julian.

Evans-Pritchard, E. E. 1965. *Theories of Primitive Religion*. London: Oxford University Press.

Ewin, D. M. 1984. "Hypnosis in Surgery and Anesthesia." Pp. 210–35 in *Clinical Hypnosis, A Multidisciplinary Approach*, ed., W. C. Wester and A. H. Smith. Philadelphia: J. B. Lippincott.

———. 1986. "Emergency Room Hypnosis for the Burned Patient." *American Journal of Clinical Hypnosis* 29: 7–12.

Facchinetti, F., M. L. Matteo, G. P. Artini, A. Volpe, A. R. Genazzani. 1997. "An Increased Vulnerability to Stress Is Associated with a Poor Outcome of In Vitro Fertilization-embryo Transfer Treatment." *Fertility and Sterility* 67: 309–14.

Fate Magazine. 1997. *Psychic Pets and Spirit Animals: True Stories from the Files of Fate Magazine*. St. Paul, Minn.: Llewellyn.

Ferreira, Antonio J. 1965. "Emotional Factors in the Prenatal Environment: A Review." *Journal of Nervous and Mental Disease* 141: 108–18.

Finkler, Kaja. 1985. *Spiritualist Healers in Mexico: Successes and Failures of Alternative Therapeutics*. New York: Bergin and Garvey.

Foldes, Jochanan J. 1975. "Ovulatory Disturbances of Psychosomatic Origin." Pp. 330–33 in *The Family*, ed. Herman Hirsch. Fourth International Congress of Psychosomatic Obstetrics and Gynecology. Basel; New York: Karger.

Fox, John W. 1992. "The Structure, Stability, and Social Antecedents of Reported Paranormal Experiences." *Sociological Analysis* 53: 417–31.

Fox, M. W. 1974. *Concepts in Ethology*. Minneapolis: University of Minnesota Press.

Frank, Jerome. 1973. *Persuasion and Healing*. Baltimore: Johns Hopkins University Press.

Fredericks, L. 1978. "Teaching Hypnosis in the Overall Approach to the Surgical Patient." *American Journal of Clinical Hypnosis* 20: 175–83.

Freud, Sigmund. [1927] 1964. *The Future of an Illusion.* Translated by W. D. Robson-Scott, revised and edited by James Strachey. Garden City, N.Y.: Doubleday. Original edition, *Die Zukunft einer Illusion,* Vienna.

Freund, Peter E. S., and Meredith B. McGuire. 1995. *Health, Illness, and the Social Body: A Critical Sociology,* 2nd ed. Englewood Cliffs, N.J.: Prentice Hall.

Fromm, Erika. 1987. "Significant Developments in Clinical Hypnosis during the Past 25 Years." *International Journal of Clinical and Experimental Hypnosis* 35: 215–30.

Fromm, Erika, and Michael R. Nash, eds. 1992. *Contemporary Hypnosis Research.* New York: Guilford.

Gaddis, Vincent, and Margaret Gaddis. 1970. *The Strange World of Animals and Pets.* New York: Cowles.

Gallagher, Bernard J., III, with Corinne J. Rita. 1995. *The Sociology of Mental Illness,* 3rd ed. Englewood Cliffs, N.J.: Prentice Hall.

Gallup, George, Jr., and F. Newport. 1991. "Belief in Paranormal Phenomena among Adult Americans." *Skeptical Inquirer* 15: 137–47.

Garrison, Vivian. 1977. "The 'Puerto Rican Syndrome' in Psychiatry and Espiritismo." Pp. 383–449 in *Case Studies in Spirit Possession,* ed. Vincent Crapanzano and Vivian Garrison. New York: John Wiley and Sons.

Gauld, Alan. 1968. *The Founders of Psychical Research.* New York: Schocken.

Gauld, Alan, and A. D. Cornell. 1979. *Poltergeists.* London: Routledge and Kegan Paul.

Geertz, Clifford, 1966. "Religion as a Cultural System." Pp. 1–46 in *Anthropological Approaches to the Study of Religion,* ed. Michael Banton. London: Tavistock.

Gibbons, Don, and James de Jarnette. 1972. "Hypnotic Susceptibility and Religious Experience." *Journal for the Scientific Study of Religion* 11: 152–56.

Gibbons, F. X., C. S. Carver, M. F. Scheier, and S. E. Hormuth. 1979. "Self-Focused Attention and the Placebo Effect: Fooling Some of the People Some of the Time." *Journal of Experimental Social Psychology* 15: 263–74.

Gibson, H. B. 1985. "Dreaming and Hypnotic Susceptibility: A Pilot Study." *Perceptual and Motor Skills* 60: 387–94.

Glaser, B. G., and A. Strauss. 1967. *The Discovery of Grounded Theory.* Chicago: Aldine.

Glik, Deborah C. 1986. "Psychosocial Wellness among Participants in Spiritual Healing Groups: A Comparison Groups Survey." *Social Science and Medicine* 22: 579–86.

Glock, Charles, and Rodney Stark. 1965. *Religion and Society in Tension.* Skokie, Ill.: Rand McNally.

Goldenweiser, A. [1917] 1965. "Religion and Society: A Critique of Émile Durkheim's Theory of the Origin and Nature of Religion." Reprint, pp. 204–16 in *Reader in Comparative Religion,* ed. W. Lessa and E. Vogt. New York: Harper and Row.

Goodall, Jane. 1971. *In the Shadow of Man.* Boston: Houghton Mifflin.

———. 1975. "The Chimpanzee." Pp. 131–70 in *The Quest for Man,* ed. Vanne Goodall. New York: Praeger.

———. 1986. *The Chimpanzees of Gombe: Patterns of Behavior.* Cambridge: The Belknap Press of Harvard University Press.

Gorsuch, R. L., and M. K. Key. 1974. "Abnormalities of Pregnancy as a Function of Anxiety and Life Stress." *Psychosomatic Medicine* 36: 352–61.

Greeley, Andrew M. 1975. *Sociology of the Paranormal: A Reconnaissance.* Beverly Hills, Calif.: Sage.

———. 1987. "Mysticism Goes Mainstream." *American Health* 6: 47–49.

Greenleaf, M., S. Fisher, C. Miaskowski, and R. DuHamel. 1992. "Hypnotizability and Recovery from Cardiac Surgery." *American Journal of Clinical Hypnosis* 35: 119–28.

Grindal, Bruce. 1983. "Into the Heart of Sisala Experience: Witnessing Death Divination." *Journal of Anthropological Research* 39, no. 1: 60–80.

Gurney, Edmund, Frederic W. H. Myers, and Frank Podmore. [1886] 1970. *Phantasms of the Living*. 2 vols. Reprint, Gainsville, Fla.: Scholars' Facsimiles and Reprints. Facsimile reproduction.

Guthrie, Stewart. 1993. *Faces in the Clouds: A New Theory of Religion*. New York: Oxford University Press.

———. 1997. "McClenon's 'Shamanic Healing, Human Evolution, and the Origin of Religion': A Critique." *Journal for the Scientific Study of Religion* 36: 355–57.

Haber, Suzanne N., and Patricia Barchas. 1984. "The Regulatory Effect of Social Rank on Behavior after Amphetamine Administration." Pp. 119–32 in *Social Hierarchies: Essays toward a Sociophysiological Perspective*, ed. P. Barchas and S. P. Mendoza. Westport, Conn.: Greenwood.

Haraldsson, Erlendur. 1985. "Representative National Surveys of Psychic Phenomena: Iceland, Great Britain, Sweden, USA, and Gallup's Multinational Survey." *Journal of the Society for Psychical Research* 53: 145–58.

Haraldsson, Erlendur, and J. M. Houtkooper. 1991. "Psychic Experience in the Multinational Human Values Study: Who Reports Them." *Journal of the American Society for Psychical Research* 85: 145–65.

Hardy, Alister. 1970. "A Scientist Looks at Religion." *Proceedings of the Royal Institute of Great Britain* 43: 201.

———. 1979. *The Spiritual Nature of Man: A Study of Contemporary Religious Experience*. Oxford: Clarendon.

Harmon, T. M., M. T. Hynan, and T. E. Tyre. 1990. "Improved Obstetric Outcomes Using Hypnotic Analgesia and Skill Mastery Combined with Childbirth Education." *Journal of Consulting and Clinical Psychology* 58, no. 5: 525–30.

Harner, Michael J. 1973. *Hallucinogens and Shamanism*. London: Oxford University Press.

Harris, Marvin. 1997. *Culture, People, Nature: An Introduction to General Anthropology*, 7th ed. New York: Longman.

Hart, Hornell Norris. 1959. *The Enigma of Survival: The Case for and against an After Life*. Springfield, Ill.: Charles C Thomas.

Hart, R. 1980. "The Influence of a Taped Hypnotic Induction Treatment Procedure on the Recovery of Surgery Patients." *International Journal of Clinical and Experimental Hypnosis* 28: 324–32.

Hartmann, Ernest. 1991. *Boundaries in the Mind: A New Psychology of Personality*. New York: Basic Books.

Hathaway, D. 1986. "Effect of Preoperative Instruction on Postoperative Outcomes: A Meta-Analysis." *Nursing Research* 35: 269–75.

Haviland, William A. 1997. *Anthropology*, 8th ed. Fort Worth, Tex.: Harcourt Brace.

Hay, David. 1979. "Religious Experiences amongst a Group of Post-Graduate Students: A Qualitative Study." *Journal for the Scientific Study of Religion* 18: 164–82.

———. 1985. "Religious Experience and Its Induction." Pp. 135–50 in *Advances in the Psychology of Religion*, ed. L. B. Brown. Oxford: Pergamon.

———. 1990. *Religious Experience Today: Studying the Facts*. London: Mowbray.

Hay, David, and Morisy, Ann. 1978. "Reports of Ecstatic, Paranormal, or Religious Experience in Great Britain and the United States—A Comparison of Trends. *Journal for the Scientific Study of Religion* 17, no. 3: 255–68.

Haynes, Renée. 1972. *The Hidden Springs: An Inquiry into Extra-Sensory Perception.* Boston: Little, Brown.

Heide, Frederick J., W. L. Wadlington, and Richard M. Lundy. 1980. "Hypnotic Responsivity as a Predictor of Outcome in Meditation." *International Journal of Clinical and Experimental Hypnosis* 28: 358–85.

Heinze, Ruth-Inge. 1990. *Shamans of the 20th Century.* New York: Irvington.

Henry, J. P., and P. M. Stephens. 1977. *Stress, Health, and the Social Environment.* New York: Springer-Verlag.

Hilgard, Ernest R. 1977. *Divided Consciousness: Multiple Controls in Human Thought and Action.* New York: Wiley.

———. 1992. "Dissociation and Theories of Hypnosis." Pp. 69–101 in *Contemporary Hypnosis Research,* ed. Erika Fromm and Michael R. Nash. New York: Guilford.

Hilgard, Ernest R., and Josephine R. Hilgard. 1983. *Hypnosis in the Relief of Pain,* 2nd ed. Los Altos, Calif.: William Kaufmann.

Hilgard, Ernest R., and Charles T. Tart. 1966. "Responsiveness to Suggestions Following Waking and Imagination Instructions and Following Induction of Hypnosis." *Journal of Abnormal Psychology* 71: 196–208.

Hillig, J. A., and J. Holroyd, 1997/1998. "Consciousness, Attention, and Hypnoidal Effects during Firewalking." *Imagination, Cognition, and Personality* 17: 153–64.

Hobbes, Thomas. [1651] 1991. *Leviathan.* Reprint, edited by Richard Tuck. New York: Cambridge University Press.

Hobson, J. Allan. 1994. *The Chemistry of Conscious States: How the Brain Changes Its Mind.* Boston: Little, Brown.

Homer. 1963. *The Odyssey.* Translated by Robert Fitzgerald. Garden City, N.Y.: Doubleday.

Honorton, Charles. 1977. "Psi and Internal Attention States." Pp. 435–72 in *Handbook of Parapsychology,* ed. Benjamin Wolman. New York: Van Nostrand Reinhold.

Hood, Ralph W. 1973. "Hypnotic Susceptibility and Reported Religious Experience." *Psychological Reports* 33: 549–50.

———, ed. 1995. *Handbook of Religious Experience.* Birmingham, Ala.: Religious Education Press.

Hopkins, Mildred B., Jeanette M. Jordan, and Richard M. Lundy. 1991. "The Effects on Hypnosis and Imagery on Bleeding Time: A Brief Communication." *International Journal of Clinical and Experimental Hypnosis* 39: 134–39.

Horton, Robin. 1967. "African Traditional Thought and Western Science." *Africa* 37, no. 3: 50–71.

Hoskovec, J., and D. Svorad. 1969. "The Relationship between Human and Animal Hypnosis." *American Journal of Clinical Hypnosis* 11: 180–82.

Hufford, David J. 1982. *The Terror That Comes in the Night: An Experience-Centered Study of Supernatural Assault Traditions.* Philadelphia: University of Pennsylvania Press.

———. 1983. "Folk Healers." Pp. 306–19 in *Handbook of American Folklore,* ed. Richard M. Dorson. Bloomington: Indiana University Press.

———. 1985. "Commentary: Mystical Experience in the Modern World." Pp. 87–183 in *The World Was Flooded with Light: A Mystical Experience Remembered,* by G. W. Foster. Pittsburgh: University of Pittsburgh Press.

———. 1988. "Contemporary Folk Medicine." Pp. 228–64 in *Unorthodox Medicine in America,* ed. Norman Gevitz. Baltimore: Johns Hopkins University Press.

———. 1993. "Epistemologies of Religious Healing." *The Journal of Philosophy and Medicine* 18: 173–92.

Hultkrantz, Ake. 1992. *Shamanic Healing and Ritual Drama.* New York: Crossroads.

Hume, David. [1748] 1988. *An Enquiry concerning Human Understanding*. Reprint, Buffalo, N.Y.: Prometheus Books.

Irwin, Harvey. 1994. *An Introduction to Parapsychology*, 2nd ed. Jefferson, N.C.: McFarland.

James, W. H. 1963. "Control Data for Evaluating the Efficacy of Psychotherapy in Habitual Spontaneous Abortion." *British Journal of Psychiatry* 109: 81–83.

James, William. [1902] 1985. *The Varieties of Religious Experience: A Study of Human Nature*. Cambridge: Harvard University Press. Original edition, New York: Longman.

Jemmott, John B., III, and Steven E. Locke. 1984. "Psychosocial Factors, Immune Mediation, and Human Susceptibility to Infectious Disease: How Much Do We Know? *Psychological Bulletin* 95: 78–108.

Jenkins, M. W., and M. H. Pritchard. 1993. "Hypnosis: Practical Applications and Theoretical Considerations in Normal Labor." *British Journal of Obstetrics and Gynecology* 100, no. 3: 221–26.

Jolly, Clifford, and Randall White. 1995. *Physical Anthropology and Archaeology*. New York: McGraw-Hill.

Juznic, N., L. Vojvodic, and D. Avramovic. 1979. "Psychoprophylaxis Today. Evaluation of Psychological Analgesia during Delivery." Pp. 951–54 in *Emotion and Reproduction*, ed. L. Carenza and L. Zichella. Fifth International Congress of Psychosomatic Obstetrics and Gynecology. London: Academic Press.

Kalweit, Holger. 1992. *Shamans, Healers, and Medicine Men*, translated by Michael H. Kohn. Boston: Shambhala.

Kane, Stephen M. 1982. "Holiness Ritual Fire Handling: Ethnographic and Psychological Considerations." *Ethos* 10: 369–84.

Kapferer, Bruce. 1983. *A Celebration of Demons: Exorcism and the Aesthetics of Healing in Sri Lanka*. Bloomington: Indiana University Press.

Karp, Ivan. 1989. "Power and Capacity in Rituals of Possession." Pp. 91–109 in *Creativity of Power: Cosmology and Action in African Societies*, ed. W. Arens and Ivan Karp, Washington, D.C.: Smithsonian Institution Press.

Katz, Richard. 1976. "Education for Transcendence: !Kia-Healing with the Kalahari !Kung." Pp. 282–301 in *Kalahari Hunter-Gatherers: Studies of the !Kung San and Their Neighbors*, ed. Richard B. Lee and Irven De Vore. Cambridge: Harvard University Press.

———. 1982. *Boiling Energy: Community Healing among the Kalahari Kung*. Cambridge: Harvard University Press.

Kellehear, Allan. 1996. *Experiences Near Death: Beyond Medicine and Religion*. New York: Oxford University Press.

Kim, Wonsik. 1967. "Korean Shamanism and Hypnosis." *American Journal of Clinical Hypnosis* 9: 193–97.

King, M. 1985. *Being Pakeha: An Encounter with New Zealand and the Maori Renaissance*. Auckland, New Zealand: Hodder & Stoughton.

King, Sallie B. 1988. "Two Epistemological Models for the Interpretation of Mysticism." *Journal of the American Academy of Religion* 56: 257–79.

Kirsch, Irving. 1990. *Changing Expectations: A Key to Effective Psychotherapy*. Pacific Grove, Calif.: Brooks/Cole.

Kirsch, Irving, and James R. Council. 1992. "Situational and Personality Correlates of Hypnotic Responsiveness." Pp. 267–91 in *Contemporary Hypnosis Research*, ed. Erika Fromm and Michael R. Nash. New York: Guilford.

Kleegman, Sophia J., and Sherwin A. Kaufman 1966. *Infertility in Women*. Philadelphia: Davis.

Kleinman, Arthur. 1980. *Patients and Healers in the Context of Culture: An Exploration of the Borderland between Anthropology, Medicine, and Psychiatry.* Berkeley, Calif.: University of California Press.

Kleinman, Arthur, and L. H. Sung. 1979. "Why Do Indigenous Practitioners Successfully Heal? A Follow-up Study of Indigenous Practice in Taiwan." *Social Science and Medicine* 13B: 7–26.

Kliment, V. 1979. "Some Psychosomatic Problems in Obstetrics." Pp. 805–8 in *Emotion and Reproduction*, Fifth International Congress of Psychosomatic Obstetrics and Gynecology, eds. L. Carenza and L. Zichella. London: Academic Press.

Köhler, Wolfgang. 1927. *The Mentality of Apes*, 2nd ed., translated by Ella Winter. New York: Harcourt, Brace.

Kok, L. P. 1989. "Hypnotic Susceptibility in Kavadi Carriers in Singapore." *Annals of the Academy of Medicine, Singapore* 18, no. 6: 655–57.

Kornhauser, Richard H., R. R. Springman, A. R. Bodenheimer, and B. Lunefeld. 1975. "Family Aspects of Brief Psychotherapy of Secondary Male Impotence." Pp. 96–99 in *The Family*, ed. Herman Hirsch. Fourth International Congress of Psychosomatic Obstetrics and Gynecology. Basel; New York: Karger.

Kouretas, Demetrios. 1967a. "Amphiareion, A Precursor of the Aesculapian Temples of Ancient Greece. *Bulletin of the Menninger Clinic* 31: 129–35.

———. 1967b. "The Oracle of Trophonius: A Kind of Shock Treatment Associated with Sensory Deprivation in Ancient Greece." *British Journal of Psychiatry* 113: 1441–46.

Koval, Joel. 1990. "Beyond the Future of an Illusion: Further Reflections on Freud and Religion." *Psychoanalytic Review* 77: 69–87.

Krippner, Stanley. 1988. "'Energy Medicine' in Indigenous Healing Systems." Pp. 191–202 in *Proceedings of the Fourth International Conference on the Study of Shamanism and Alternate Modes of Healing*. Madison, Wisc.: A-R Editions.

Krippner, Stanley, and Jeanne Achterberg. 2000. "Anomalous Healing Experiences." Pp. 353–95 in *Varieties of Anomalous Experience: Examining the Scientific Evidence*, ed. Etzel Cardeña, Steven Jay Lynn, and Stanley Krippner. Washington, D.C.: American Psychological Association.

Krippner, Stanley, and Leonard George. 1986. "Psi Phenomena as Related to Altered States of Consciousness." Pp. 332–64 in *Handbook of States of Consciousness*, ed. Benjamin B. Wolman and Montague Ullman. New York: Van Nostrand Reinhold.

Kuhn, Thomas S. 1970. *The Structure of Scientific Revolutions*, 2nd ed. Chicago: University of Chicago Press.

Kurtz, Paul. 1985. *A Skeptic's Handbook of Parapsychology.* Buffalo, N.Y.: Prometheus.

Laderman, Carol. 1987. "The Ambiguity of Symbols in the Structure of Healing." *Social Science and Medicine* 24: 293–301.

———. 1991. *Taming the Wind of Desire: Psychology, Medicine, and Aesthetics in Malay Shamanistic Performance.* Berkeley, Calif.: University of California Press.

Lambo, T. Adeoye. 1974. "Psychotherapy in Africa." *Psychotherapy and Psychosomatics* 24: 311–26.

Lang, Elvira V., Eric G. Benotsch, Lauri J. Fick, Susan Lutgendorf, Michael L. Berbaum, Kevin S. Berbaum, Henrietta Logan, and David Spiegel. 2000. "Adjuctive Non-Pharmacological Analgesia for Invasive Medical Procedures: A Randomized Trial." *Lancet* 355: 1486–90.

Larson, David B., and Susan S. Larson. 1994. *The Forgotten Factor in Physical and Mental Health: What Does the Research Show?* Rockville, Md.: National Institute for Healthcare Research.

Larson, David B., K. A. Sherrill, J. S. Lyons, F. C. Craigie, S. B. Thielman, M. A. Greenwold, S. S. Larson. 1992. "Dimensions and Valences of Measures of Religious Commitment Found in the American Journal of Psychiatry and the Archives of General Psychiatry: 1978 through 1989." *American Journal of Psychiatry* 149: 557–59.

Laughlin, Charles D., Jr., John McManus, Eugene G. d'Aquili. 1992. *Brain, Symbol, and Experience: Toward a Neurophenomenology of Human Consciousness.* New York: Columbia University Press.

Lawlis, G. F., D. Selby, D. Hinnant, E. McCoy. 1985. "Reduction of Postoperative Pain Parameters by Presurgical Relaxation Instructions for Spinal Pain Patients." *Spine* 10, no. 7: 649–51.

Lazarus, Richard, Allen D. Kanner, and Susan Folkman. 1980. "Emotions: A Cognitive-Phenomenological Analysis." Pp. 189–217 in *Emotion: Theory, Research, and Experience*, Vol. 1, ed. Robert Plutchik and Henry Killerman. New York: Academic Press/Harcourt Brace Jovanovich.

Leavitt, Judith Walzer. 1986. *Brought to Bed: Childbearing in America, 1750–1950.* New York: Oxford University Press.

Leikind, Bernard J., and William J. McCarthy. 1985. "An Investigation of Firewalking." *Skeptical Inquirer* 10, no. 1: 23–34.

Lewin, Roger. 1991. "Stone Age Psychedelia." *New Scientist* 130 (8 June): 30–34.

Lewis, I. M. 1971. *Ecstatic Religion: An Anthropological Study of Spirit Possession and Shamanism.* Middlesex, U.K.: Penguin.

Lewis-Williams, J. David, and Thomas A. Dowson. 1988. "The Signs of All Times: Entoptic Phenomena in Upper Paleolithic Art." *Current Anthropology* 29: 201–45.

Lex, Barbara. 1979. "The Neurobiology of Ritual Trance." Pp. 117–51 in *The Spectrum of Ritual: A Biogenetic Structural Analysis* ed. Eugene G. d'Aquili, C.D. Laughlin Jr., and John McManus. New York: Columbia University Press.

Lieberman, Philip. 1984. *The Biology and Evolution of Language.* Cambridge: Harvard University Press.

———. 1997. *Eve Spoke: Human Language and Human Evolution.* New York: W. W. Norton.

Lobel, Marci. 1994. "Conceptualizations, Measurement, and Effects of Prenatal Maternal Stress on Birth Outcomes." *Journal of Behavioral Medicine* 19: 225–72.

Long, Joseph K., ed. 1977. *Extrasensory Ecology: Parapsychology and Anthropology.* Metuchen, N.J.: Scarecrow Press.

Lorenz, Konrad. 1966. *On Aggression,* translated by M. Wilson. New York: Harcourt, Brace, and World.

Ludwig, Arnold. 1966. "Altered States of Consciousness." *Archives of General Psychiatry* 15: 225–34.

Lumsden, Charles L., and Edward O. Wilson. 1983. *Promethean Fire: Reflections on the Origin of Mind.* Cambridge: Harvard University Press.

McCaffery, M., and A. Beebe. 1989. *Pain: Clinical Manual for Nursing Practice.* St. Louis: C.V. Mosby.

McClenon, James. 1984. *Deviant Science: The Case of Parapsychology.* Philadelphia: University of Pennsylvania Press.

———. 1991. "Near-Death Folklore in Medieval China and Japan: A Comparative Analysis." *Asian Folklore Studies* 50: 319–42.

———. 1993. "Surveys of Anomalous Experience in Chinese, Japanese, and American Samples." *Sociology of Religion* 54: 295–302.

———. 1994. *Wondrous Events: Foundations of Religious Belief.* Philadelphia: University of Pennsylvania Press.

———. 1996. "The Experience of Care Project: Students as Participant Observers in the Hospital Setting." *Academic Medicine* 71: 923–29.

———. 1997a. "Shamanic Healing, Human Evolution, and the Origin of Religion." *Journal for the Scientific Study of Religion* 36, no. 3: 345–54.

———.1997b. "Spiritual Healing and Folklore Research: Evaluating the Hypnosis/Placebo Theory." *Alternative Therapies in Health and Medicine* 3, no. 1: 61–66.

———. 2000. "Content Analysis of an Anomalous Memorate Collection: Testing Hypotheses Regarding Universal Features." *Sociology of Religion* 61, no. 2: 155–69.

McDonald, R. L. 1968. "The Role of Emotional Factors in Obstetrical Complications: A Review." *Psychosomatic Medicine* 30: 222.

McFalls, Joseph A., Jr. 1979. *Psychopathology and Subfecundity.* New York: Academic Press.

Mac Hovec, Frank J. 1975. "Hypnosis before Mesmer." *American Journal of Clinical Hypnosis* 17: 215–20.

———. 1976. "The Evil Eye: Superstition of Hypnotic Phenomenon?" *American Journal of Clinical Hypnosis* 19: 74–79.

———. 1979. "The Cult of Asklipios." *American Journal of Clinical Hypnosis* 22: 85–90.

Mackett, John. 1989. "Chinese Hypnosis." *British Journal of Experimental and Clinical Hypnosis* 6: 129–30.

MacLean, Paul D. 1973. "A Triune Concept of Brain and Behavior." Pp. 6–66 in *The Hinks Memorial Lectures,* ed. T. Boag and D. Campbell. Toronto: Toronto University Press.

Majno, Guido. 1975. *The Healing Hand: Man and Wound in the Ancient World.* Cambridge: Harvard University Press.

Makarec, Katherine, and Michael A. Persinger. 1990. "Electroencephalographic Validation of a Temporal Lobe Signs Inventory in a Normal Population." *Journal of Research in Personality* 24: 323–37.

Malan, J. 1932. "The Possible Origin of Religion as a Conditioned Reflex." *American Mercury* 25: 314–17.

Malinowski, Bronislaw. [1931] 1965. "The Role of Magic and Religion." Reprint, pp. 102–11 in *Reader in Comparative Religion,* 2nd ed., ed. William A. Lessa and Evon Z. Vogt. New York: Harper and Row.

Mandell, Arnold J. 1980. "Toward a Psychobiology of Transcendence: God in the Brain." Pp. 379–464 in *The Psychobiology of Consciousness,* ed. Julian M. Davidson and Richard J. Davidson. New York: Plenum.

Mann, Edward C. 1959. "Habitual Abortion." *American Journal of Obstetrics and Gynecology* 77: 706–18.

Marcuse, F. L. 1951. "Individual Differences in Animal Hypnosis." *British Journal of Medical Hypnosis* 3: 17–20.

———. 1961. "Interpretation in Animal Hypnotism." *Personality* 1: 240–42.

———, ed. 1964. *Hypnosis throughout the World.* Springfield, Ill.: Charles C. Thomas.

Margolis, Clorinda G., Barbara B. Domangue, Carole Ehleben, and Linda Shrier. 1983. "Hypnosis in the Early Treatment of Burns: A Pilot Study." *American Journal of Clinical Hypnosis* 26: 9–15.

Marmer, M. J. 1959. "Hypnoanalgesia and Hypnoanesthesia for Cardiac Surgery." *Journal of the American Medical Association* 171: 512–17.

Marshall, Gary D., and Philip G. Zimbardo. 1979. "Affective Consequences of Inadequately Explained Physiological Arousal." *Journal of Personality and Social Psychology* 37: 970–88.

Maslach, Christina. 1979. "Negative Emotional Biasing of Unexplained Arousal." *Journal of Personality and Social Psychology* 37: 953–69.

Masters, William H., and Virginia E. Johnson. 1970. *Human Sexual Inadequacy.* Boston: Little, Brown.

Mehl-Madrona [Mehl], Lewis E. 1994. "Hypnosis and Conversion of the Breech to the Vertex Presentation." *Archives of Family Medicine* 3: 881–87.

———. 1997. "Lessons in Coyote Medicine." *Shaman's Drum* March/May: 27–33.

Moerman, Daniel E. 1979. "Anthropology of Symbolic Healing." *Current Anthropology* 20: 59–80.

Molinski, Hans. 1975. "Different Behavior of Women in Labor as a Symptom of Different Psychic Patterns." Pp. 338–42 in *The Family,* ed. Herman Hirsch. Fourth International Congress of Psychosomatic Obstetrics and Gynecology. Basel; New York: Karger.

Moody, Raymond A., Jr. 1975. *Life after Life.* Atlanta, Ga.: Mockingbird.

Moore, Lawrence Earle, and Jerold Zelig Kaplan. 1983. "Hypnotically Accelerated Wound Healing." *American Journal of Clinical Hypnosis* 26: 16–19.

Morgan, Arlene H. 1973. "The Heritability of Hypnotic Susceptibility in Twins." *Journal of Abnormal and Social Psychology* 82: 55–61.

Morgan, Arlene H., and Ernest R. Hilgard. 1973. "Age Differences in Susceptibility to Hypnosis." *International Journal of Clinical and Experimental Hypnosis* 21: 78–85.

Morgan, Arlene H., Ernest R. Hilgard, and Edna C. Davert. 1970. "The Heritability of Hypnotic Susceptibility of Twins: A Preliminary Report." *Behavioral Genetics* 1: 213–24.

Morris, Brian. 1987. *Anthropological Studies of Religion: An Introductory Text.* Cambridge: Cambridge University Press.

Morris, Norman. 1983. "Labor." Pp. 281–308 in *Handbook of Psychosomatic Obstetrics and Gynecology,* ed. Lorraine Dennerstein and Graham D. Burrows. Amsterdam: Elseveir Biomedical Press.

Müller, F. Max. 1889. *Natural Religion.* London: Longmans.

Mullings, Leith. 1984. *Therapy, Ideology and Social Change: Mental Healing in Urban Ghana.* Berkeley, Calif.: University of California Press.

Mumford, E., H. Schlesinger, and G. Glass. 1982. "The Effects of Psychological Intervention on Recovery from Surgery and Heart Attacks: An Analysis of the Literature." *American Journal of Public Health* 72, no. 2: 141–51.

Murdock, George Peter. 1945. "The Common Denominator of Cultures." Pp. 123–42. in *The Science of Man in the World Crisis,* ed. Ralph Linton. New York: Columbia University Press.

Murdock, George Peter, and D. White. 1969. "Standard Cross-Cultural Sample." *Ethnology* 8: 329–69.

Murphy, Michael. 1992. *The Future of the Body: Explorations into the Future Evolution of Human Nature.* Los Angeles: Jeremy P. Tarcher.

Nace, Edgar P., Arthur M. Warwick, Ronald L. Kelley, and Frederick J. Evans. 1982. "Hypnotizability and Outcome in Brief Psychotherapy." *Journal of Clinical Psychiatry* 43: 129–33.

Nadon, Robert, and John F. Kihlstrom. 1987. "Hypnosis, Psi, and the Psychology of Anomalous Experience." *Behavioral and Brain Sciences* 10: 597–99.

Needham, Rodney. 1967. "Percussion and Transition." *Man* 2: 606–14.

Neher, Andrew. 1962. "A Physiological Explanation of Unusual Behavior in Ceremonies Involving Drums." *Human Biology* 34: 151–61.

Neppe, Vernon M. 1983. "Temporal Lobe Symptomatology in Subjective Paranormal Experiences." *Journal of the American Society for Psychical Research* 77: 1–31.

Nichols, L. A. 1968. "Correspondence (The Oracle of Trophonius)." *British Journal of Psychiatry* 114: 650–51.

Noyes, R. W., and E. M. Chapnick. 1964. "Literature on Psychology and Infertility: A Critical Analysis." *Fertility and Sterility* 15: 543–58.

Omer, Haim, Zvi Palti, and Dov Friedlander. 1986. "Evaluating Treatments of Preterm Labor: Possible Solutions for Some Methodological Problems." *European Journal of Obstetrics and Reproductive Biology* 22: 229–36.

Orne, Martin T. 1977. "The Construct of Hypnosis: Implications of the Definition for Research and Practice." *Annals of the New York Academy of Sciences* 296: 14–33.

Ortiz de Montellano, Bernard. 1975. "Empirical Aztec Medicine." *Science* 188: 215–20.

Otto, Rudolph. 1931. *The Philosophy of Religion*. London: Williams and Norgate.

Owen, Iris M., and Margaret Sparrow. 1976. *Conjuring Up Philip: An Adventure in Psychokinesis*. New York: Harper and Row.

Pahnke, Walter N. 1966. "Drugs and Mysticism." *International Journal of Parapsychology* 8: 295–320.

———. 1967. "The Mystical and/or Religious Element in the Psychedelic Experience." Pp. 41–56 in *Do Psychedelics Have Religious Implications?* ed. D. H. Salman and R. H. Prince. Montreal: R. M. Bucks Memorial Society. (Reprinted, with modifications, in *Clinical Psychiatry and Religion*, ed. E. M. Pattison. Boston: Little, Brown, 1969, pp. 149–62.)

Pahnke, Walter N., and W. A. Richards. 1966. "Implications of LSD and Experimental Mysticism." *Journal of Religion and Health* 5: 175–208.

Palmer, John. 1979. "A Community Mail Survey of Psychic Experiences." *Journal of the American Society for Psychical Research* 73: 221–51.

Parker, Adrian. 1975. *States of Mind: ESP and Altered States of Consciousness*. New York: Taplinger.

Parker, C. 1978. "The Relative Importance of Group Selection in the Evolution of Primate Societies." Pp. 103–12 in *Human Behavior and Adaptation*, ed. N. Blurton Jones and V. Reynolds. New York: Halsted.

Patterson, D. R., G. L. Burns, J. J. Everett, and J. A. Marvin. 1992. "Hypnosis for the Treatment of Burn Pain." *Journal of Consulting and Clinical Psychology* 60, no. 5: 713–17.

Pekala, Ronald J., and B. Ersek. 1992/1993. "Firewalking versus Hypnosis: A Preliminary Study Concerning Consciousness, Attention, and Fire Immunity." *Imagination, Cognition, and Personality* 12: 207–29.

Pekala, Ronald J., V. K. Kumar, and James Cummings. 1992. "Types of High Hypnotically Susceptible Individuals and Reported Attitudes and Experiences of the Paranormal and Anomalous." *Journal of the American Society for Psychical Research* 86: 135–50.

Perry, Campbell, Robert Nadon, and Jennifer Button. 1992. "The Measurement of Hypnotic Ability." Pp. 459–90 in *Contemporary Hypnosis Research*, ed. Erika Fromm and Michael R. Nash. New York: Guilford.

Persinger, Michael A. 1984a. "People Who Report Religious Experiences May Display Enhanced Temporal Lobe Signs." *Perceptual and Motor Skills* 58: 963–75.

———. 1984b. "Propensity to Report Paranormal Experiences Is Correlated with Temporal Lobe Signs." *Perceptual and Motor Skills* 59: 583–86.

Persinger, Michael A., and Katherine Makarec. 1987. "Temporal Lobe Epileptic Signs and Correlative Behaviors Displayed by Normal Populations." *Journal of General Psychology* 114: 179–95.

———. 1993. "Complex Partial Epileptic Signs as a Continuum from Normals to

Epileptics: Normative Data and Clinical Populations." *Journal of Clinical Psychology* 49: 33–45.

Persinger, Michael A., and P. M. Valliant. 1985. "Temporal Lobe Signs and Reports of Subjective Paranormal Experiences in a Normal Population: A Replication." *Perceptual and Motor Skills* 60: 903–9.

Pert, Candace B. 1997. *Molecules of Emotion: The Science behind Mind-Body Medicine.* New York: Touchstone.

Pfeiffer, John E. 1982. *The Creative Explosion: An Inquiry into the Origins of Art and Religion.* New York: Harper and Row.

Piccione, Carlo, Ernest R. Hilgard, and Philip G. Zimbardo. 1989. "On the Degree of Stability of Measured Hypnotizability over a 25-Year Period. *Journal of Personality and Social Psychology* 56: 289–95.

Preus, J. Samuel. 1987. *Explaining Religion: Criticism and Theory from Bodin to Freud.* New Haven: Yale University Press.

Price, Harry. 1936. *Bulletin II: A Report on Two Successful Firewalks.* London: London Council for Psychical Investigation.

———. 1937. "Firewalking." *Nature* 139 (29 May): 928–29.

Prill, H. J. 1983. "Methods in Birth Preparation." Pp. 269–80 in *Handbook of Psychosomatic Obstetrics and Gynecology,* Lorraine Dennerstein and Graham D. Burrows. Amsterdam: Elsevier Biomedical Press.

Prioreschi, Plinio. 1992. "Supernatural Elements in Hippocratic Medicine." *Journal of the History of Medicine* 47: 389–404.

Proudfoot, Wayne. 1985. *Religious Experience.* Berkeley: University of California Press.

Putnam, John J. 1988. "The Search for Modern Humans." *National Geographic* 174: 439–77.

Radcliffe-Brown, Alfred Reginald. 1922. *The Andaman Islanders.* Cambridge: Cambridge University Press.

———. [1939] 1965. "Taboo." Reprint, pp. 112–23 in *Reader in Comparative Religion,* 2nd ed., ed. William A. Lessa and Evon Z. Vogt. New York: Harper and Row.

Ramachandran, Vilayanur S., and Sandra Blakeslee. 1998. *Phantoms in the Brain: Probing the Mysteries of the Human Mind.* New York: William Morrow and Company.

Ramachandran, Vilayanur S., W. S. Hirstein, K. C. Armel, E. Tecoma, and V. Iragui. 1997. "The Neural Basis of Religious Experience." *Society for Neuroscience Abstracts* 23 (part 2): 1316.

Rao, K. Ramakrishna, and John Palmer. 1987. "The Anomaly Called Psi: Recent Research and Criticism." *Behavioral and Brain Sciences* 10: 539–51, 634–43.

Rapkin, D., M. Straubing, and J. Holroyd. 1991. "Guided Imagery, Hypnosis and Recovery from Head and Neck Cancer Surgery: An Exploratory Study." *International Journal of Clinical and Experimental Hypnosis* 39: 4.

Rappaport, Roy A. 1979. *Ecology, Meaning, and Religion.* Berkeley: North Atlantic Books.

Rauscher, E. A., and B. A. Rubik. 1980. "Effects on Motality Behavior and Growth of Salmonella Typhimurium in the Presence of a Psychic Subject." Pp. 140–42 in *Research in Parapsychology 1979,* ed. W. G. Roll. Metuchen, N.J.: Scarecrow Press.

———. 1983. "Human Volitional Effects on a Model Bacterial System." *Psi Research* 2: 38–48.

Rawlings, R. M. 1972. "The Inheritance of Hypnotic Amnesia." Paper presented at the 44th Congress of the Australian and New Zealand Society for the Advancement of Science, July 1972. Cited by Morgan (1973).

Redekop, Calvin. 1967. "Toward an Understanding of Religion and Social Solidarity." *Sociological Analysis* 27: 149–61.

Reisenzein, Rainer. 1983. "The Schachter Theory of Emotion: Two Decades Later." *Psychological Bulletin* 94: 239–64.

Reiser, O. L. 1932. "The Biological Origins of Religion." *Psychoanalytic Review* 19: 1–22.

Rennie, Bryan. Forthcoming. "Mircea Eliade: A Secular Mystic in the History of Religions?" In *Method as Path: Religious Experience and Hermeneutical Reflection,* ed. Elliot R. Wolfson and Jeffrey J. Kripal. New York University Annual Conference in Comparitive Religion. New York: Seven Bridges Press.

Reynolds, Vernon, and R. E. S. Tanner. 1995. *The Social Ecology of Religion.* Oxford: Oxford University Press. Original edition, 1983, *The Biology of Religion,* New York: Longman.

Rhine, Louisa E. 1951. "Conviction and Associated Conditions in Spontaneous Cases." *Journal of Parapsychology* 15: 164–91.

———. 1954. "Frequency of Types of Experience in Spontaneous Precognition." *Journal of Parapsychology* 18: 93–123.

———. 1981. *The Invisible Picture: A Study of Psychic Experiences.* Jefferson, N.C.: McFarland and Company.

Richards, Douglas G. 1990. "Hypnotic Susceptibility and Subjective Psychic Experiences." *Journal of Parapsychology* 54: 35–51.

Richards, John Thomas. 1982. *SORRAT: A History of the Neihardt Psychokinesis Experiments, 1961–1981.* Metuchen, N.J.: Scarecrow Press.

Richards, P., and Michael A. Persinger. 1991. "Temporal Lobe Signs, the Dissociative Experience Scale, and Hemisphericity." *Perceptual and Motor Skills* 72: 1139–42.

Richeport, Madeleine M. 1992. "The Interface between Multiple Personality, Spirit Mediumship, and Hypnosis." *American Journal of Clinical Hypnosis* 34: 168–77.

Ring, Kenneth. 1980. *Life at Death: A Scientific Investigation of the Near-Death Experience.* New York: Coward, McCann and Geoghegan.

Rock, John A. 1986. "Investigation and Management of Habitual Abortion." Pp. 275–302 in *Reproductive Failure,* ed. Alan H. DeChemey. New York: Churchill Livingstone.

Rogers, M., and Reich, P. 1986. "Psychological Intervention and Surgical Patients: Evaluation Outcome." *Advances in Psychosomatic Medicine* 15: 25–50.

Rogo, D. Scott. 1975. *Parapsychology: A Century of Inquiry.* New York: Dell.

———. 1982. *ESP and Your Pet.* New York: Grossett and Dunlap.

Ross, J., and Michael A. Persinger. 1987. "Positive Correlations between Temporal Lobe Signs and Hypnosis Induction Profiles: A Replication." *Perceptual and Motor Skills* 64: 828–30.

Rossi, E. L. 1986. *The Psychobiology of Mind-Body Healing.* New York: Norton.

Runciman, W. G. 1970. *Sociology in Its Place.* Cambridge: Cambridge University Press.

Rutherford, R. 1965. "Emotional Aspects of Infertility." *Clinical Obstetrics and Gynecology* 8: 100–14.

Sabom, Michael. 1982. *Recollections of Death: A Medical Investigation.* New York: Simon and Schuster.

Sammons, Robert. 1992. "Parallels between Magico-Religious Healing and Clinical Hypnosis Therapy." Pp. 53–67 in *Herbal and Magical Medicine: Traditional Healing Today,* ed. James Kirkland, Holly F. Mathews, C. W. Sullivan III, and Karen Baldwin. Durham, N.C.: Duke University Press.

Sanders, Raymond S., and Joseph Reyher. 1969. "Sensory Deprivation and the Enhancement of Hypnotic Susceptibility." *Journal of Abnormal Psychology* 74: 375–81.

Sapolsky, Robert M. 1990. "Stress in the Wild." *Scientific American* 262: 116–23.

Schachter, Stanley, and Jerome E. Singer. 1962. "Cognitive, Social, and Physiological Determinants of Emotional State." *Psychological Review* 69: 379–99.

Schechter, E. I. 1984. "Hypnotic Induction vs. Control Conditions: Illustrating an Approach to the Evaluation of Replicability in Parapsychology." *Journal of the American Society for Psychical Research* 78: 1–27.

Schleiermacher, Friedrich. [1799] 1996. *On Religion: Speeches to Its Cultured Despisers.* Reprint, translated and edited by Richard Crouter. New York: Cambridge University Press.

Schouten, Syro A. 1979. "Analysis of Spontaneous Cases as Reported in 'Phantasms of the Living.'" *European Journal of Parapsychology* 2, no. 4: 408–55.

———. 1981. "Analyzing Spontaneous Cases: A Replication Based on the Sannwald Collection." *European Journal of Parapsychology* 4, no. 1: 9–48.

———. 1982. "Analyzing Spontaneous Cases: A Replication Based on the Rhine Collection." *European Journal of Parapsychology* 4, no. 2: 113–58.

Schul, Bill. 1977. *The Psychic Power of Animals.* Greenwich, Conn.: Fawcett.

Schumaker, John F. 1990. *Wings of Illusion.* Buffalo, N.Y.: Prometheus.

———. 1995. *The Corruption of Reality: A Unified Theory of Religion, Hypnosis, and Psychopathology.* Amherst, N.Y.: Prometheus.

Seibel, Machelle M., and Melvin L. Taymor. 1982. Emotional Aspects of Infertility. *Fertility and Sterility* 37: 137–45.

Shames, Victor A., and Patricia G. Bowers. 1992. "Hypnosis and Creativity." Pp. 334–63 in *Contemporary Hypnosis Research,* ed. Erika Fromm and Michael R. Nash. New York: Guilford.

Sharon, Douglas. 1978. *Wizard of the Four Winds: A Shaman's Story.* New York: Free Press.

Sheehan, Peter. W. 1992. "The Phenomenology of Hypnosis and the Experiential Analysis Technique." Pp. 364–89 in *Contemporary Hypnosis Research,* ed. Erika Fromm and Michael R. Nash. New York: Guilford.

Sheldrake, Rupert. 1995. *Seven Experiments That Could Change the World: A Do-It-Yourself Guide to Revolutionary Science.* New York: Riverhead Books.

Shively, C. A., K. Laber-Laird, and R. F. Anton. 1997. "Behavior and Physiology of Social Stress and Depression in Female Cynomolgus Monkeys." *Biological Psychiatry* 41, no. 8: 871–82.

Shor, Ronald E., and Emily Carota Orne. 1962. *Harvard Group Scale of Hypnotic Susceptibility.* Palo Alto, Calif.: Consulting Psychologists Press.

Shostak, Marjorie. 1981. *Nisa: The Life and Words of a !Kung Woman.* Cambridge: Harvard University Press.

Sidgwick, Henry, and Committee. 1894. "Report on the Census of Hallucinations." *Proceedings of the Society for Psychical Research* 10: 25–422.

Siegel, R. K., and M. E. Jarvik. 1975. "Drug-induced Hallucinations in Animals and Man." Pp. 81–161 in *Hallucinations: Behavior, Experience, and Theory,* ed. R. K. Siegel and L. J. West. New York: Wiley.

Sigerist, Henry E. [1951] 1987a. *A History of Medicine.* Vol. 1, *Primitive and Archaic Medicine.* New York: Oxford University Press. Original edition, New York: Oxford University Press.

———. [1961] 1987b. *A History of Medicine.* Vol. 2, *Early Greek, Hindu, and Persian Medicine.* New York: Oxford University Press. Original edition, New York: Oxford University Press.

Smirnov, Yuri. 1989. "Intentional Human Burial: Middle Paleolithic (Last Glaciation) Beginnings." *Journal of World Prehistory* 3: 199–233.

Snow, Loudell. 1993. *Walkin' over Medicine.* Boulder: Westview.

Southam, A. L. 1960. "What to Do with the 'Normal' Infertile Couple." *Fertility and Sterility* 11: 543–48.

Spencer, Herbert. [1876] 1969. *The Principles of Sociology,* ed. Stanislav Andreski. Hamden, Conn.: Archon. Original edition, London: Williams and Norgate.

Spiegel, David, and L. H. Albert. 1983. "Naloxone Fails to Reverse Hypnotic Alleviation of Chronic Pain." *Psychopharmacology* 81: 140–43.

Spiegel, Herbert, and David Spiegel. 1987. *Trance and Treatment: Clinical Uses of Hypnosis,* 2nd ed. New York: Basic. (1st ed., 1978.)

Stahler, F., E. Stahler, and R. Gutanian. 1972. "Perinatal Mortality of the Child Lowered by Psychoprophylaxis." P. 56 in *Psychosomatic Medicine, Obstetrics, and Gynecology,* ed. Herman Hirsch. Basel; New York: Karger. (Cited by Juznic, Vojvodic, and Avramovic, 1979.)

Stam, Henderikus J., and Nicholas P. Spanos. 1982. "The Asclepian Dream Healings and Hypnosis: A Critique." *International Journal of Clinical and Experimental Hypnosis* 30: 9–22.

Swatos, William H., Jr., and Loftur Reimar Gissurarson. 1997. *Icelandic Spiritualism: Mediumship and Modernity in Iceland.* New Brunswick, N.J.: Transaction Publishers.

Swirsky-Sacchetti, Thomas, and Clorinda G. Margolis. 1986. "The Effects of a Comprehensive Self-Hypnosis Training Program on the Use of Factor VIII in Severe Hemophilia." *International Journal of Clinical and Experimental Hypnosis* 34: 71–83.

Syme, S. L. 1975. "Social and Psychological Risk Factors in Coronary Heart Disease." *Modern Concepts of Cardiovascular Disease* 44: 17–21.

Syrjala, K. L., C. Cummings, G. W. Donaldson. 1992. "Hypnosis or Cognitive Behavioral Training for the Reduction of Pain and Nausea During Cancer Treatment: A Controlled Clinical Trial." *Pain* 48: 137–46.

Targ, Elisabeth, Marilyn Schlitz, and Harvey J. Irwin. 2000. "Psi-Related Experiences." Pp. 219–52 in *Varieties of Anomalous Experience: Examining the Scientific Evidence,* ed., Etzel Cardeña, Steven Jay Lynn, and Stanley Krippner. Washington, D.C.: American Psychological Association.

Tart, Charles T., ed. 1969. *Altered States of Consciousness.* New York: John Wiley and Sons.

Taussig, Michael. 1987. *Shamanism, Colonialism, and the Wild Man: A Study in Terror and Healing.* Chicago: University of Chicago Press.

Tellegen, A., and G. Atkinson. 1974. "Openness to Absorbing and Self-Altering Experiences ('Absorption'), a Trait Related to Hypnotic Susceptibility." *Journal of Abnormal Psychology* 83: 268–77.

Templeton, A. A., and G. L. Penney. 1982. "The Incidence, Characteristics, and Prognosis of Patients Whose Infertility Is Unexplained." *Fertility and Sterility* 37: 175–82.

Tennant, C. 1988. "Psychosocial Causes of Duodenal Ulcer." *Australian and New Zealand Journal of Psychiatry* 22: 195–201.

Thalbourne, Michael A., Luciana Bartemucci, Peter S. Delin, Bronwyn Fox, and Oriana Nofi. 1997. "Transliminality: Its Nature and Correlates." *Journal of the American Society for Psychical Research* 91: 305–31.

Thalbourne, Michael A., and Peter S. Delin. 1994. "A Common Thread Underlying Belief in the Paranormal, Creative Personality, Mystical Experience, and Psychopathology." *Journal of Parapsychology* 58: 3–38.

————. 1999. "Transliminality: Its Relation to Dream Life, Religiosity, and Mystical Experience." *International Journal for the Psychology of Religion* 9: 35–43.

Theorell, T., and R. H. Rahe. 1975. "Life Change Events, Ballistocardiography and Coronary Death." *Journal of Human Stress* 1: 18–24.

Thomas, L. E., and P. E. Cooper. 1980. "Incidence and Psychological Correlates of Intense Spiritual Experience." *Journal of Transpersonal Psychology* 12: 75–85.

Thompson, Stith. 1966. *Motif-Index of Folk-Literature*. 6 vols. Bloomington: Indiana University Press.

Tiller, S. G., and Michael A. Persinger. 1994. "Elevated Incidence of a Sensed Presence and Sexual Arousal During Partial Sensory Deprivation and Sensitivity to Hypnosis: Implications for Hemisphericity and Gender Differences." *Perceptual and Motor Skills* 79: 1527–31.

Trinkaus, Erik. 1983. *The Shanidar Neanderthals*. New York: Academic Press.

Trivers, Robert. 1985. *Social Evolution*. Menlo Park, Calif.: Benjamin/Cummings.

Tseng, Wen-Shing. 1975. "The Nature of Somatic Complaints among Psychiatric Patients: The Chinese Case." *Comprehensive Psychiatry* 16: 237–45.

Tupper, Carl, and Robert J. Weil. 1962. "The Problem of Spontaneous Abortion." *American Journal of Obstetrics and Gynecology* 83: 421–24.

————. 1968. "The Etiology of Habitual Abortion." Pp. 751–65 in *Progress in Infertility*, ed. S. J. Behrman and Robert W. Kistner. Boston: Little Brown.

Turner, Edith. 1992. *Experiencing Ritual: A New Interpretation of African Healing*. Philadelphia: University of Pennsylvania Press.

————. 1993. "The Reality of Spirits: A Tabooed or Permitted Field of Study?" *Anthropology of Consciousness* 4: 9–12.

Tylor, Edward Burnett. [1871] 1958. *Religion in Primitive Culture*. New York: Harper and Brothers. Original edition, *Primitive Culture*, London: John Murray.

Tyrrell, G. N. M. [1942] 1963. *Apparitions*. Reprint, New York: Collier.

Ullman, M. 1947. "Herpes Simplex and Second Degree Burn Induced under Hypnosis." *American Journal of Psychiatry* 103: 823–30.

Utian, Wulf H., James M. Goldfarb, and Miriam B. Rosenthal. 1983. "Psychological Aspects of Infertility." Pp. 231–48 in *Handbook of Psychosomatic Obstetrics and Gynecology*, ed. Lorraine Dennerstein and Graham D. Burrows. Amsterdam: Elsevier Biomedical Press.

Van der Does, A. J. W., and R. Van Dyck. 1989. "The Effectiveness of Standardized versus Individualized Hypnotic Suggestions: A Brief Communication." *International Journal of Clinical and Experimental Hypnosis* 27: 1–5.

Van Dyck, Richard, and Kees Hoogduin. 1990. "Hypnosis: Placebo or Nonplacebo?" *American Journal of Psychotherapy* 44: 396–404.

Vaneechoutte, Mario, and John R. Skoyles. 1998. "The Memetic Origin of Language: Modern Humans as Musical Primates." *Journal of Memetics: Evolutionary Models of Information Transmission* 2. Available: <http://www.cpm.mmu.ac.uk/jom-emit/1998/vol2/vaneechoutte_m&skoyles_jr.html>.

Van Lawick-Goodall, Jane. 1968. *The Behavior of Free-living Chimpanzees in the Gombe Stream Reserve*. London: Baillière, Tindall & Cassell.

Vialou, Denis. 1998. *Prehistoric Art and Civilization*, translated by Paul G. Bahn. New York: Harry N. Abrams.

Virtanen, Leea. 1990. *"That Must Have Been ESP!"* translated by John Atkinson and Thomas Dubois. Bloomington, Ind.: Indiana University Press. Original edition, *Telepaattiset Kokemukset*, Helsinki: Werner Söderström Osakeyhtiö, 1977.

Vogel, Virgil. 1970. *American Indian Medicine.* New York: Ballantine.

Völgyesi, Ferenc András. 1966. *Hypnosis of Man and Animals,* 2nd ed., revised in collaboration with Gerhard Klumbies. Baltimore: Williams and Wilkins.

Wadhwa, Pathik D., Christine Dunkel-Schetter, Aleksandra Chicz-DeMet, Manuel Porto, and Curt A. Sandman. 1996. "Prenatal Psychosocial Factors and the Neuroendocrine Axis in Human Pregnancy." *Psychosomatic Medicine* 58: 432–46.

Wagner, M. W., and F. H. Ratzeburg. 1987. "Hypnotic Suggestibility and Paranormal Belief." *Psychological Reports* 60: 1,069–70.

Walker, Alan. 1993. *The Narioktome Homo Erectus Skeleton.* Cambridge: Harvard University Press.

Walker, Jearl. 1977. "The Amateur Scientist." *Scientific American* 237, no. 2 (August): 126–31.

Walker, Priscilla Campbell, and R. F. Q. Johnson. 1974. "The Influence of Presleep Suggestions on Dream Content: Evidence and Methodological Problems." *Psychological Bulletin* 81: 362–70.

Wallace, Anthony. 1966. *Religion: An Anthropological View.* New York: Random House.

Waller, Niels G., Brian A. Kojetin, Thomas J. Bouchard Jr., David T. Lykken, and Auke Tellegen. 1990. "Genetic and Environmental Influences on Religious Interests, Attitudes, and Values: A Study of Twins Reared Apart and Together." *Psychological Science* 1: 138–42.

Warner, Richard. 1985. *Recovery from Schizoprenia: Psychiatry and Political Economy.* London: Routledge and Kegan Paul.

Wax, Murray L. 1984. "Religion as Universal: Tribulations of an Anthropological Enterprise." *Zygon: Journal of Religion and Science* 19: 5–20.

Weiner, Herbert. 1992. *Perturbing the Organism: The Biology of Stressful Experience.* Chicago: University of Chicago Press.

Weitzenhoffer, A. M., and B. M. Sjoberg. 1961. "Suggestibility with and without 'Induction of Hypnosis.'" *Journal of Nervous and Mental Disease* 13: 205–20.

West, D. K. 1948. "A Mass-Observational Questionnaire on Hallucinations." *Journal of the Society for Psychical Research* 34: 187–96.

West, Donald J. 1990. "A Pilot Census of Hallucinations." *Proceedings of the Society for Psychical Research* 57 (part 215): 163–207.

White, R. 1985. "Thoughts on Social Relationships and Language in Hominid Evolution." *Journal of Social and Personal Relationships* 2: 95–115.

Wickramasekera, Ian E. 1977. "On Attempts to Modify Hypnotic Susceptibility: Some Psychophysiological Procedures and Promising Directions." *Annals of the New York Academy of Sciences* 296: 143–53.

———. 1987. "Risk Factors Leading to Chronic Stress-Related Symptoms." *Advances: Journal of the Institute for the Advancement of Health* 4: 9–35.

———. 1988. *Clinical Behavioral Medicine: Some Concepts and Procedures.* New York: Plenum.

Wilkinson, J. C. M., B. A. Lieberman, Elizabeth M. Belsey, and R. W. Beard. 1975. "Psychological Profiles of Women Complaining of Pelvic Pain." Pp. 241–45 in *The Family,* ed. Herman Hirsch. Fourth International Congress of Psychosomatic Obstetrics and Gynecology. Basel; New York: Karger.

Wilson, David Sloan, and Elliott Sober. 1994. "Reintroducing Group Selection to the Human Behavioral Sciences." *Behavioral and Brain Sciences* 17: 585–654.

Wilson, Sheryl C., and Theodore X. Barber. 1978. "The Creative Imagination Scale as a Measure of Hypnotic Responsiveness: Applications to Experimental and Clinical Hypnosis." *American Journal of Clinical Hypnosis* 20: 235–49.

———. 1983. "The Fantasy-Prone Personality: Implications for Understanding Imagery, Hypnosis, and Parapsychological Phenomena." Pp. 340–87 in *Imagery: Current Theory, Research, and Application*, ed. Anees A. Sheikh. New York: John Wiley and Sons.

Winkelman, Michael. 1982. "Magic: A Theoretical Reassessment." *Current Anthropology* 23: 37–66.

———. 1986. "Trance States: A Theoretical Model and Cross-Cultural Analysis." *Ethos* 14: 174–203.

———. 1992. *Shamans, Priests, and Witches: A Cross-Cultural Study of Magico-Religious Practitioners*. Tempe: Arizona State University Anthropological Research Papers, no. 44.

———. 1997. Personal communication based on analysis of stratified subsample of 45 societies selected from the Standard Cross-Cultural Sample. (Cited in Murdock and White, 1969).

———. 2000. *Shamanism: The Neural Ecology of Consciousness and Healing*. Westport, Conn.: Bergin and Garvey.

Woody, Erik Z., Kenneth S. Bowers, and Jonathan M. Oakman. 1992. "A Conceptual Analysis of Hypnotic Responsiveness: Experience, Individual Differences, and Context." Pp. 3–33 in *Contemporary Hypnosis Research*, ed. Erika Fromm and Michael R. Nash. New York: Guilford.

Worrall, Ambrose A., with Olga N. Worrall. 1970. *The Gift of Healing: A Personal Story of Spiritual Therapy*. New York: Harper and Row.

Wright, Robert. 1994. *The Moral Animal: Evolutionary Psychology and Everyday Life*. New York: Vantage.

Wulff, David M. 1997. *Psychology of Religion: Classic and Contemporary*, 2nd ed., New York: John Wiley and Sons.

Wylder, Joseph Edward. 1978. *Psychic Pets*, New York: Stonehill.

Xiao, E., and M. Ferin. 1997. Stress-related disturbances of the Menstrual Cycle. *Annals of Medicine* 29: 215–19.

Young, David, and Jean-Guy Goulet, eds. 1994. *Being Changed by Cross-Cultural Experience: The Anthropology of Extraordinary Experience*. Peterborough, Ontario: Broadview.

Zaleski, Carol. 1987. *Otherworld Journeys: Accounts of Near-Death Experiences in Medieval and Modern Times*. New York: Oxford University Press.

Zichella, L., A. Luchetti, D. Salmaggi, A. Bartoleschi, M. Sprecavisciole, M. Vinciguerra, P. Pietrobattista, P. Pancheri, A. Connolly, and O. Sachetti. 1979. "Psychological and Psychoneuroendocrine Correlates of Normal Labor." Pp. 731–39 in *Emotion and Reproduction*, Fifth International Congress of Psychosomatic Obstetrics and Gynecology, ed. L. Carenza and L. Zichella. London: Academic Press.

Zigler-Shani, Z., D. A. Kipper, D. M. Serr, V. Insler, and G. Oelsner. 1975. "Psychological Characteristics of Women with Functional Infertility." Pp. 315–18 in *The Family*, ed. Herman Hirsch. Fourth International Congress of Psychosomatic Obstetrics and Gynecology. Basel; New York: Karger.

Zimmer, E. Z., B. A. Peretz, E. Eyal, and K. Fuchs. 1988. "The Influence of Maternal Hypnosis on Fetal Movements in Anxious Pregnant Women." *European Journal of Obstetrics and Reproductive Biology* 27: 133–37.

Index